D1625669

Half-Earth Socialism

Half-Earth Socialism

*A Plan to Save the Future
from Extinction, Climate
Change, and Pandemics*

Troy Vettese and
Drew Pendergrass

VERSO

London • New York

First published by Verso 2022
© Troy Vettese and Drew Pendergrass 2022

1 3 5 7 9 10 8 6 4 2

Verso
UK: 6 Meard Street, London W1F 0EG
US: 20 Jay Street, Suite 1010, Brooklyn, NY 11201
versobooks.com

Verso is the imprint of New Left Books

ISBN-13: 978-1-83976-031-0
ISBN-13: 978-1-83976-032-7 (UK EBK)
ISBN-13: 978-1-83976-033-4 (US EBK)

British Library Cataloguing in Publication Data
A catalogue record for this book is available from the British Library

Library of Congress Cataloging-in-Publication Data
A catalog record for this book is available from the Library of Congress

Typeset in Sabon by MJ & N Gavan, Truro, Cornwall
Printed in the UK by CPI Group

To our parents

Contents

A map of the world that does not include Utopia is not worth even glancing at, for it leaves out the one country at which Humanity is always landing. And when Humanity lands there, it looks out, and, seeing a better country, sets sail. Progress is the realisation of Utopias.

–Oscar Wilde

We enter into Utopia's proper and new-found space: the education of desire. This is not the same as 'a moral education' towards a given end: it is, rather, to open a way to aspiration, to 'teach desire to desire, to desire better, to desire more, and above all to desire in a different way'.

–E. P. Thompson,
quoting Miguel Abensour

Introduction

Is it so incomprehensible that the people today cry out for utopias, for powerful presentations of their future fate?

–Otto Neurath

Looking Backward: 2047

In the autumn of 2029, after many years of ravaging the cities and hamlets of poor nations, climate change proved itself capable of bringing even the heartland of global capitalism to its knees. Swollen by the unseasonably warm waters in the north-west Atlantic, a hurricane of unprecedented ferocity left an arc of destruction from Washington, DC, to Boston.[1] Powerful storm surges deluged coastal towns and strong winds downed power lines, leaving 30 million people in darkness for weeks. As emergency crews dug through the rubble, even the most fanatical Republicans could no longer deny the effects of climate change. A consensus, reached in a candle-lit session of Congress, was not to decarbonize the energy system, but rather to deploy a radical and untested technology called solar radiation management (SRM) to ensure such a calamity would never again befall the United States.

The government contracted a start-up, spun from an Ivy League laboratory, to douse the heavens with a sulphuric mist. High-flying military jets were retrofitted to dump a payload of atmosphere-altering sulphur into the stratosphere. The resulting 'stratoshield' of reflective aerosols blocked out the sun by a carefully calibrated fraction and reduced global temperatures to pre-industrial levels within a few years. Respectable opinion conceded that while it was tragic that SRM caused a slew of poor harvests in equatorial countries and the additional atmospheric sulphur killed thousands of people every year, on the whole, the benefits surely exceeded the costs.[2] Rather than seeing SRM as a dangerous and desperate measure, optimists portrayed it as a demonstration of American statesmanship, technology, and entrepreneurial pluck.

Soon, however, the costs of the SRM programme became impossible to overlook. A pernicious development was the sulphuric aerosols' steady erosion of the ozone layer – a protective shield upon which all earthly life depends. The geoengineers assured the public that an ozone-neutral aerosol would soon be found. They experimented with diamonds and engineered nanoparticles, and for a time they were especially excited by calcium carbonate because its alkalinity appeared capable of reversing the ozone layer's acidification.[3] Unfortunately, the complex chemistry of the atmosphere meant that the calcium carbonate unexpectedly catalysed a reaction that actually left the ozone hole bigger than before.[4] By the 2040s – more than a decade into the SRM programme – there was still no long-term solution to the problem. At this point SRM could not simply be switched off, because the high concentrations of greenhouse gases would heat the atmosphere all at once in what scientists called 'termination shock'.

While the threat to the ozone layer lingered on the horizon, SRM's disruption of various global weather systems was a clear and present danger. The most worrying was the weakening of the monsoon, which threatened the livelihood of tens of millions of Indian farmers. Through diplomacy and generous restitution, Washington managed to talk Delhi out of its threats to shoot down the American SRM fleet, but it was uncertain whether a similar agreement could be brokered with Moscow or Beijing if those governments confronted an SRM-induced disaster. Washington, however, cared little about what non-nuclear powers thought of the stratoshield, having implemented it roughshod over objections from other countries in 2029. American unilateralism in SRM research dated back to the late 2010s, when a coalition of African and low-lying island nations repeatedly tried to bring SRM under an international authority, such as the UN Environment Programme or the Montreal Protocol (a treaty which protects the ozone layer). The US had vetoed these motions to keep SRM unregulated; geoengineering, it seems, had always been a form of planetary class war.[5]

To ward off accusations of climate imperialism, geoengineers claimed that SRM was actually in the interest of poor nations.[6] SRM, according to this argument, lowered poor countries' risk premiums for 'catastrophe bonds', an exotic financial instrument hawked by Wall Street bankers keen to greenwash their portfolios. In this way, the geoengineers believed, the market could bridge the divide that separated the Global North and South. There was an opportunity to test this market solution soon after the 2029 deployment began, when unprecedented droughts wracked West Africa. Yet these crises usually did not meet all the conditions laid out in some contracts, leaving cash-strapped governments struggling to respond.[7] Even when bondholders did pay out, the money often came too late to aid relief efforts, nor could it buy back ecosystems that had deteriorated under the new SRM regime.[8] Such experiences contrasted sharply with SRM's impacts in the core capitalist states, where quotidian life continued more or less as normal save for the nearly permanent overcast weather. Even then, many saw blue skies as an inevitable casualty of modernity, much like electrification's extinction of starry nights a century before.

SRM marked the beginning of the end for the environmental movement. With chemicals partially blocking the energy source for solar panels, investors panicked and funds for renewable infrastructure crashed in the early 2030s, sparking an unexpected renaissance for the high-cost, environmentally destructive 'nonconventional' oil sector – tar sands, fracking, and deep-sea rigs. Indeed, far from being curtailed, total petroleum production was on track to reach 116 million barrels per day by 2040, some 16 per cent more than in the early 2020s.[9] With the stratoshield in place, the imperative to abolish the fossil-fuel industry slackened. While SRM returned a measure of climatic and economic stability (if only in the rich North), this new global thermostat proved unable to reverse the decline of the biosphere. The macabre drum beat of habitat loss and extinctions continued unabated. Ecologists despaired at the disappearance of countless species whose life cycles were disoriented by the syncopated seasons and shocks of freak weather. Unabated carbon pollution threw off the ocean's chemistry to the point where only the hardiest creatures could survive in the vast acidic wastes. Sulphuric aerosols created acid rain that poisoned forests and lakes, undoing one of the great triumphs of environmental activism during the 1980s.

In sum, these events spelled a strange defeat for the environmental movement – strange because for decades it had won victory after victory. With millions demonstrating in the streets for climate justice in the 2010s, environmentalist parties took power in regional and national governments around the world in the 2020s, allowing them to finally realize their dream of 'green capitalism'. For example, carbon pricing, which covered only a fifth of global emissions in 2020, increased to half by 2030.[10] Unfortunately, the median price only rose from US $15/tonne to $40 (translating to a mere $0.36 a gallon at the pump). This fell well short of the more stringent targets set by the Intergovernmental Panel on Climate Change (IPCC), ranging from $135 to $6,050/tonne (i.e., topping out at an extra $53.24 per gallon of gasoline).[11] The greens were more successful in implementing new global standards that ended up doubling the rate of energy efficiency growth between 2017 and 2040. Yet, such improvements were counteracted by total energy demand growing even faster.[12] Relative gains matter little on a finite planet. Proponents of 'green' cars (electric, fuel-cell, or hybrid) faced a similar set of contradictory trends. These vehicles made up a fifth of the global fleet in 2040 and 30 per cent of new sales – a real achievement – but because people were buying and driving cars at higher rates than ever, the total amount of oil guzzled by personal transportation barely budged.[13] One major reason for this was the failure of green governments to reduce the demand for cars through increased urban density and public transportation. In 2040, wind and solar only made up 4 per cent of the energy system despite being the fastest-growing sources of power generation, while fossil fuels maintained a diminished but still commanding 76 per cent share.[14] The problem was that the greens mistook slowing down the pace of the environmental crisis for victory, rather than merely a defeat postponed.

After decades of environmentalists' championing 'win–win' solutions for both business and nature, it became clear that making unprofitable decisions was where true freedom lay. The 'free' market forbade shutting down fossil-fuel firms, implementing energy caps, and building large-scale renewable-energy infrastructure. Private utilities fiercely resisted the latter because they dreaded the renewable energy–induced 'death spiral': when too many people installed their own solar panels, utilities lost customers and were forced to raise prices, which in turn led to further shrinking of their market share. What's more, these new consumer-producers destabilized the grid by selling excess energy on windy or sunny days. Utilities responded by lobbying hard against 'feed-in tariffs' and licences for renewable energy production.[15] Even if governments managed to overcome such resistance, the variability of wind and sun coupled with insufficient energy storage meant that disruptions in the energy supply were inevitable.[16] Imposing such inconveniences was political suicide in the

Global North, even if brown-outs had long been common in the South.[17] The whole premise of 'green capitalism' was that environmentalists would only make minimal demands of firms and consumers in order to gain their support – but how could the world's greatest problem be solved by such modest means?

Such political reticence extended, with perhaps the direst consequences, to the meat question. Environmentalists had long been loath to raise it in fear of losing support, but this proved a grave miscalculation. While ocean acidification from carbon pollution and the new SRM programme pushed many species to extinction, the greatest butcher of global biodiversity was the livestock industry.[18] Despite constituting only a few percentage points of global GDP, animal husbandry ravaged countless wild ecosystems to sustain captive life in its teeming billions. Meat production doubled over the three decades before 2047, with devastating costs to local environments and the global climate.[19] This future was supposed to have been averted by entrepreneurial scientists and ethical firms purveying 'clean meat' (lab-grown or plant-based), but while this new market grew significantly, just as with electric cars and the renewable energy sector, it could not solve the problem by itself. The market could sell both the poison and its antidote, but it cared little about the right ratio of the two.

As a planetary force comparable to the fossil-fuel industry, the livestock sector generated repeated shocks in the world system over these bleak decades. Million-animal operations were hothouses of zoonotic illnesses, and small-scale outbreaks occurred almost constantly: *E. coli* (including the dangerous STEC O104:H4 strain), Q fever, and salmonella contaminated water, air, and food.[20] However, these crises were mere pinpricks compared with the civilizational laceration of the avian flu pandemic of 2035, some three decades after the first instance of human-to-human transmission. Given that the virus' victims suffered a mortality rate of 60 per cent, containing the global death toll to only 200 million seemed a pyrrhic

victory of sorts.[21] After this annus horribilis, there were calls
to drain disease reservoirs within wild animal populations
through intentional extinctions.[22] This was seen as more expedient than asking people to give up meat and expand nature
preserves to act as cordons sanitaires, although public health
experts had been advocating such a programme since the early
twenty-first century.[23]

It was difficult to prise the environmental and economic
catastrophes apart during these years. The inexorable rise
of factory farming wiped out the remnants of the world's
10,000-year-old peasantry. With little industry to absorb this
displaced class, the share of humanity living in slums more
than doubled between the early 2020s and 2047 to 3 billion
people.[24] Inequality, automation, and low rates of economic
growth meant that by 2040, some 24 per cent of the world's
population was reduced to involuntary indolence, a fourfold
increase compared with the mid-2010s.[25] By 2050, the richest
1 per cent had funnelled 39 per cent of the world's wealth
into their pockets, dwarfing the 27 per cent held by the global
middle class (i.e., the middle two-fifths of humanity), let alone
the scraps held by the bottom billions.[26] Inequality had environmental consequences too, as the top 1 per cent emitted
twice as much carbon as the bottom half of humanity.[27] In
the mid-2030s, the first trillionaire emerged: a Chinese tech
mogul narrowly beat an American rival to become a modern
Croesus. Geoengineering made small but still substantial fortunes for the scientist-entrepreneurs, who cashed in on the IPO
of their start-up soon after the stratoshield was put in place.
Conspiracy theorists who saw SRM as poisonous 'chemtrails'
ensured that the geoengineers enjoyed little peace; some were
even assassinated.[28]

Although much of the natural world had been transformed
into a factory farm, a suburb, or a garbage dump, the market's
control over the biosphere remained far from complete. SRM
best revealed the gulf that lay between mastery and unintended

chaos. Even after years of study and implementation, the geo-engineers still hadn't fully grasped the hyper-complexity of the Earth system that spanned living creatures, the oceans' slow churn, and a vast, turbulent climate. They confronted this challenge with complacency rather than humility in the face of what they did not – and indeed could not – know. In the decades leading up to 2029, the geoengineers did not bother to collect much baseline data or build detailed models.[29] In this way, their actions belied what some philosophers of science had suspected: that small-scale SRM experiments could never capture what implementation would be like due to the complexity of the Earth system.[30] In this post-experimental era, action replaced knowledge.

In the 2030s, the material and political threat posed by climate change to the prevailing order peaked and subsided. The fact that scientist-entrepreneurs and their generous philanthropic backers overcame the climate crisis through SRM seemed to vindicate faith in the market. Fossil-fuel companies, conservative think tanks, and economics departments, after all, had been among the earliest supporters of geoengineering.[31] That conservative coalition, which had cultivated this crisis of environmental catastrophe and inequality since the mid-twentieth century, remained dominant a century later. Despite briefly tasting power, the environmentalists accomplished little because they never elucidated how the various facets of the environmental crisis – climate change, pandemics, and mass extinctions – were interlinked; nor did they articulate what a post-crisis society might actually look like. The ruling class had long been clever and ruthless, but they were also fortunate to face such hapless opponents.

The View from Mont Pèlerin

How can this dystopian future be avoided? Environmental collapse and feudal levels of inequality are not inevitable. Although the biosphere is certainly in dire shape, there is still time to reverse its decline and simultaneously create a just society. The purpose of this book is to outline the material conditions of the current ecological predicament and show how it can be transcended by providing new ways of conceiving the relationship between the economy and the environment. While at times our proposals may seem outlandish – our book, after all, belongs to the utopian tradition – they are meant to encourage those on the Left and in the environmental movement to take seriously the challenge of not merely surviving the next century but creating a better society within a wilder and stabilized biosphere.

Our thought experiment of the decades leading up to 2047 reveals the inadequacy of mainstream environmentalism. We tried to be fair by assuming the rapid uptake of carbon markets, renewable energy, and electric cars, and show how these measures would still fail to prevent a global ecocide by mid-century. It is not enough if the market for 'clean meat' or renewables grows quickly – their environmentally deleterious competitors must also contract, and this is unlikely to happen if environmental policy is guided by price signals. Indeed, our survey shows that it is not so much the monetary value or rate of growth that matters, as it is the physical composition of the global economic metabolism: How much land are we converting from forest to pasture? How much energy are we using, what are its physical properties, and how is it generated? How should we allocate necessary but environmentally costly resources such as steel and concrete?

Why, then, is politics outsourced to the market, an institution that clearly cannot address the environmental crisis? This question forces us to confront the market's high priests: the neoliberals. In 2047, they will not only celebrate the centennial of their movement's birth but likely be in a strong position to enjoy a second century of intellectual, political, and economic hegemony.

The epithet 'neoliberal' is often a grenade lobbed with the pin attached, because this explosive term is rarely understood by those hurling it. To grasp this controversial and murky ideology, it helps to return to the moment of its genesis as a self-conscious movement. On 10 April 1947, thirty-nine European and American intellectuals congregated at the Hotel du Parc, a luxurious Swiss establishment perched upon Mont Pèlerin.[32] Those attending this first meeting of the Mont Pèlerin Society – an organization that still exists – sought to reinvent liberalism for an age when the market was everywhere in retreat. The Great Depression, World War II, and the post-war welfare state made clear that classical liberalism's faith in laissez-faire was obsolete. Departing from their eighteenth-century tradition, the neoliberals recognized that markets were hardly natural but rather needed nurturing and protection by a strong state. Markets deserved such care because they could concentrate knowledge diffused throughout society into the metric of price. The conference's impresario, Friedrich Hayek, saw the price system as a mechanism not merely for exchanging goods but also 'for communicating information'.[33] Markets allowed people to act rationally as individuals without full knowledge of *why* prices change, which meant that society's 'optimal ignorance' was surprisingly high.[34]

While we disagree with the neoliberals' belief in the all-knowing market, we admire their commitment to simple and powerful axioms. If, as they claim, the market produces knowledge better than other institutions – such as science or central planning – then it follows that all of society and nature should be set to the logic of the price system by a neoliberal state. This philosophical shorthand allows neoliberals to diagnose the ills of the world and to propose a slate of prescriptions.[35] It allows them to *act*. We believe that environmentalists and socialists need a similar shorthand to regain political momentum. Thirty-four years ago, Stuart Hall proposed 'learning from Thatcherism' because neoliberals had demonstrated how 'good ideas ... don't fall off the shelf without an ideological framework to give those ideas coherence.'[36] In many ways, our political philosophy is crafted in the mirror image of neoliberalism because we similarly focus on questions of knowledge and the role of markets in society. Through this intellectual exchange, we have devised a few principles to provide the basis of what we call Half-Earth socialism.

The concept of Half-Earth comes from E. O. Wilson, an entomologist whose research has shown the need to rewild half of the planet to staunch the haemorrhaging of biodiversity. While global warming, poaching, and invasive species decimate flora and fauna, Wilson stresses that the greatest driver of extinction remains habitat loss.[37] So, why is it 'Half-Earth' and not a quarter or three-fifths? Early on in his career, Wilson and his colleague Robert MacArthur discovered a simple mathematical relationship between land area and biodiversity. In their study of island biogeography, they found that the number of species was roughly proportional to the fourth root of the area.[38] This meant that, all things being equal, there were fewer species on small islands than on large ones. Decades later, Wilson realized that nature preserves were the terrestrial equivalents of islands. As 15 per cent of the world's land area is presently protected (plus a measly 2 per cent of the ocean), only half of all species will survive the Sixth Extinction.[39] To create a global ark able to protect 84 per cent of species, then 50 per cent of Earth needs to be protected ($0.5^{0.25} = 0.84$).[40] Such costly but necessary ecological stewardship would yield many other benefits, such as sequestering atmospheric carbon in rejuvenated ecosystems and creating buffers to prevent the emergence of new zoonoses.[41] Yet Wilson fails to see that Half-Earth must be socialist if it is ever to exist. Such an enormous reform would quickly run up against entrenched economic interests, from mining firms to ranchers, many of whom would be willing to bloody their hands to protect their bottom line.[42]

As we sketch what Half-Earth socialism might look like, we strive to carefully account for what is necessary and feasible, even if such things are hardly politically expedient now. If Hayek had circumscribed his political aims with the political realities of 1947, we would not be living in a neoliberal society today. Nonetheless, our utopian imagination is constrained by quite conservative parameters derived from the scientific literature on energy production, land use, planning, and 'planetary boundaries'. Much of the book is dedicated to debunking panaceas beloved across much of the political spectrum, such as nuclear power, geoengineering, 'green growth', and carbon capture and storage. We train an equally sceptical eye on the solutions proposed by the demi-monde of Brooklyn socialists to the alpine eyrie of Davos' philanthropist kings (it's not clear why Swiss hotels are so attractive to megalomaniacs). While our book critically engages with neoliberalism, it also confronts the delusions of the political centre and Left.

This book's purpose, however, is not primarily to criticize the present but to posit a countervailing positive vision for the future. We survey the present crisis and detail how an eco-socialist alternative might work in practice. Our framework draws on ecology, energy studies, epidemiology, cybernetics, history, mathematics, climate modelling, utopian socialism, and, yes, neoliberalism. This is not to say that we have drafted the only possible solution for everything that ails the world today; this is merely a start, a provocation, for a broader but more serious discussion about life after capitalism. We want to ask the hard questions about politics in an age of ecological collapse: What is socialism? How does socialist democracy work? What does a truly environmentally stable society look like? How could an eco-socialist coalition take power? How would local, national, and global levels of government interact? Half-Earth socialism's tension between utopianism and practicality allows us to create a framework commensurate with the scale of the task at hand but is simultaneously realistic enough to provide the basis for socialism in our lifetime.

From 1989 to 2047

Neoliberal hegemony has endured so long because its opponents have repeatedly let crises go to waste. The Great Recession of 2008 left Alan Greenspan (a former Federal Reserve chairman and card-carrying member of the Mont Pèlerin Society) so disoriented that he confessed his 'mistake' in trusting the market to guide financial actors rather than relying on direct regulation.[43] The catastrophic SARS-CoV-2 pandemic was an even greater disturbance to the neoliberal order, but neither environmentalists nor socialists secured significant gains.

By contrast, the neoliberals were well prepared intellectually and organizationally to attack the welfare state when opportunities presented themselves. They took advantage of the economic instability of the 1970s to orchestrate a bloody coup in Chile (1973), followed by electoral victories in Britain (1979) and the US (1980). The neoliberals won because they paired sudden ruthlessness with a willingness to wage a decades-long war of ideas. They can be beaten, but only if socialists and environmentalists create a diverse coalition guided by shared political aims. Until that happens, the only real competition the neoliberals face will be 'racist-libertarians' – the architects of Brexit, the alt-right, and Alternative für Deutschland – in what is basically an intra-Hayekian feud rather than a real clash of ideologies.[44] If the Left and the environmentalist movement had undergone a theoretical and organizational revision of Mont Pèlerin proportions following their defeats in the 1980s, then things might not be so bleak now.

In the decades since the collapse of the Eastern bloc, the Left has lost not only time to reinvent itself politically but also crucial ecological buffers that guard against collapse. From the vantage point of 2022, the natural world of the late 1980s and early 1990s appears almost Edenic. In 1988, atmospheric carbon pollution was only a modest 350 ppm, the pie-in-the-sky target that inspires Bill McKibben's 350.org movement.[45] As we write this, 2021 is on course to be the first year in which atmospheric carbon levels have averaged 50 per cent higher (419 ppm) than the pre-industrial norm (278 ppm).[46] Well over half of all carbon emissions and most of the deforestation of the Amazon rainforest have occurred since 1990.[47] During that time, 420 million hectares of the world's forests have been razed, an area equal to India and Pakistan put together.[48] At the end of the Cold War, China had only just begun to build up what was to become its gigantic livestock industry, thus bringing its millennia-old tradition of sustainable agriculture to an end.[49] As a consequence of this agricultural industrialization, avian flu (H_5N_1) jumped for the first time from poultry to humans in 1997, with numerous outbreaks in China and elsewhere since then. The environmental crisis has *accelerated* since 1989 in large part because of the China boom – by far the largest and fastest industrial revolution ever.[50] With Deng Xiaoping's efforts to liberalize the Chinese economy in the 1980s, the fall of socialist governments, and the decimation of the global peasantry, the world market has now spread to the ends of the earth, accelerating resource extraction and leaving ecological devastation in its wake.

While time has been lost, hope need not be – if only because the neoliberals are few in number compared to those who suffer at their hands. This is not to say that unifying these multifarious millions will be easy. Making such a coalition requires that movements learn from each other and make concessions when necessary. Environmentalists must curb their Malthusianism, an ideology that blames ecological and economic problems on 'overpopulation'. A commitment to environmental justice, not the bigoted environmentalism of times past, must be central to the movement, so that people of colour – who bear the brunt of the environmental crisis – can take the lead in shaping the future. Conservationists need to work carefully with Indigenous nations to ensure that nature preserves do not continue to act as institutions of colonial exclusion.[51] Socialists need to realize that the gravity of the current crisis demands taking environmental limits seriously, even if it means giving up fantasies of a post-capitalist Cockaigne. Although intellectuals on the Left faddishly invoke 'ecology' or the 'Anthropocene', too often this is mere analytical garnish rather than rigorous engagement with contemporary science. Scientists should ally with social movements – otherwise they are doomed to model ever more unlikely climate scenarios or back foolhardy measures like SRM. The gap between socialism and science was not always so wide as it is now. In 1941, Hayek fretted that the Left 'has been strongly supported and even led by men of science and engineers'; the renewal of that alliance would strike fear in the neoliberals' coal-black hearts.[52]

The central role of the livestock industry in the climate and extinction crises means that the animal-rights movement should be a contingent within a Half-Earth socialist coalition even though they have often been the 'orphans of the Left'.[53] Vegans have often fitted awkwardly within the broader Left because of their widespread adherence to utilitarianism – a creed that uncritically accepts capitalism. Nor have vegans helped their cause with their tone-deaf comparisons of the plight of animals with marginalized groups like the disabled, Black people, and victims of the Judeocide during World War II.[54] These problems, however, aren't insurmountable and are sometimes magnified by misperceptions and clichés. In the US, for example, vegans are disproportionately working-class people of colour.[55] More than a century ago, Upton Sinclair imagined in his proletarian novel *The Jungle* that socialism would be largely vegetarian because no one should be forced to engage in the 'debasing and repulsive' work of the slaughterhouse.[56] Animal-rights groups could co-operate with workers to achieve Sinclair's aim of squeezing profit margins through line slowdowns and higher wages until the industry is abolished.

Feminists, too, would be crucial allies in this struggle. Whether we achieve Half-Earth socialism or not, changes in the labour market are already beginning to centre women workers, and will continue to do so. Jobs that are often done by women, so-called 'pink-collar' jobs in health and education, not only represent some of the strongest segments of the labour movement today but also foreshadow the shift to a zero-carbon economy that prioritizes care work over extractive labour. 'Labour movements in the nineteenth and twentieth centuries insisted that workers had built the world in the most literal sense', observes political philosopher Alyssa Battistoni. 'The labour movement of the twenty-first century needs to foreground the workers who will make it possible for us to live in it.'[57] The eco-socialist future is female. To reach such a bright horizon, however, we need to deal with the widespread misogyny in contemporary socialist, environmentalist, and animal-rights organizations.[58]

Without a shared world-view to bind this heterogeneous movement of movements, each faction risks political impotence through isolation, or worse, co-optation by the ruling neoliberal bloc. If the expansion of green infrastructure through a Green New Deal is forever postponed, it will be hard to reproach trade unions for accepting the few jobs on offer in pipeline construction. While it is necessary to end the exploitation of animals, animal-rights activists should temper their attacks on Indigenous hunting, both out of respect for a different way of life and as a matter of tactics, because native peoples have spearheaded many successful environmental campaigns.[59] Indigenous hunting, after all, is not what got us into this mess. In fact, biodiversity tends to be higher in Indigenous-managed territory than in nature preserves.[60] The pursuit of global equality, too, must be part of Half-Earth socialism if it is going to stand any chance of being realized. By refusing to countenance restrictions on energy use in the rich world, Northern environmentalists have fostered little solidarity with potential allies in the Global South.[61] And so on. Neoliberals have not needed to divide and conquer their enemies because liberatory movements have obliged with their own interminable feuds.

How such a Half-Earth socialist coalition might come to power we cannot say. In some countries it might follow a path similar to the rise of Nelson Mandela and the African National Congress in South Africa: a mixture of strikes, divestment, sabotage, elections, boycotts, and violence. In other countries a purely electoral strategy might work, but such a victory would only mark a new phase of struggle. Karl Marx approved of competing in elections but predicted that if a dedicated socialist party were to ever win, the ruling class would unleash a 'pro-slavery rebellion' against a 'peaceful and legal revolution'.[62] Nearly a century later, this prediction was borne out by the massacre in Chile in 1973. As historian Eric Hobsbawm observed in its bloody aftermath, 'the Left has generally underestimated the fear and hatred of the Right, the ease with which well-dressed men and women acquire a taste for blood'.[63] One should not expect the neoliberals to meekly accept defeat.

The Narrative Ark

In 1888, Edward Bellamy wrote *Looking Backward*, his vision of a harmonious futuristic socialism in the year 2000.[64] However, it is harder to share Bellamy's optimism when we look backward from 2047. Although the future may appear bleak now, it is all the more pressing to imagine utopian alternatives to motivate and mobilize the dispirited masses. 'To many thousands of isolated thinkers', radical economist J. A. Hobson observed, '[*Looking Backward*] offered the first distinctively moral support and stimulus to large projects of structural reform in industry and politics'.[65] For us, agreeing on the details of what that utopia might look like matters less than agreeing that speculation is a vital political act. This means reviving the utopian socialist tradition that has for far too long been marginalized by the 'scientific socialism' of Marxism. What was a nuanced critique of utopian socialism in Marx's hands became a bludgeon wielded by his epigones who scorned post-revolutionary proposals as frivolous 'recipes for the cook-shops of the future'.[66] Condescension towards utopianism is not only poetically impoverished but also a tactical mistake, because it limits the Left's ability to implement a socialist programme upon taking power. There is in fact a literal need to write recipes for society after the revolution, because the environmental crisis makes clear that those cook-shops must be vegan. Think then of *Half-Earth Socialism* as a cookbook divided into four courses: the philosophical, the material, the technical, and the imaginative.

The first chapter can be thought of as a delicate hors d'oeuvre. It attempts to set the philosophical foundations for a new eco-socialism much in the way that the neoliberals of the 1940s worked from first principles to revive liberalism. To do so we go back to 1798, when three competing philosophies of nature emerged simultaneously – those of G. W. F. Hegel, Thomas Malthus, and Edward Jenner. Hegel believed that humanity would eventually fully 'humanize' and control nature, an attitude towards nature that would later be called 'Prometheanism' and adopted by Marx and his followers. While Malthus' influence on the environmental movement peaked in the 1960s and 1970s, his dread of 'overpopulation' remains widespread among green activists and thinkers today. Jenner, who studied and popularized the smallpox vaccine, presciently warned against humanity's unnatural dominion over the animal kingdom. This survey of the late eighteenth century leads us to the central problem confronted by neoliberals during the mid-twentieth: what can we know? While neoliberals stress the unknowability of the market – which is why central planning could never replace it – we counter that nature is much more complex. The environmental crisis forces us to decide between controlling the market and controlling nature, a dilemma that is especially clear in the case of SRM. A new eco-socialism, then, must be based on the unknowability of nature and, consequently, the need to control the economy within safe limits.

The second chapter is a hearty appetizer, where we begin to use the principles developed in the first chapter to decide upon an array of solutions based on their material characteristics. We are interested in such things as yields per hectare and watts per square metre of various agricultural and energy systems. First we focus on the solutions offered by mainstream environmentalists: bioenergy carbon capture and sequestration (BECCS), nuclear power, and Wilson's Half-Earth. We show how these policies – yes, even Half-Earth – do not suffice to reverse the biosphere's deterioration. This is because energy, biodiversity, and carbon sequestration are not the three separate spheres they appear to be in environmental discourse but rather a single problem mediated through the scarcity of land. This insight, which we glean from examining the shortcomings of these three 'solutions', helps us develop the material aspects of Half-Earth socialism: veganism, renewables with energy quotas, and planetary rewilding.

The main course, the third chapter, delves into the devilishly difficult problem of planning. If it is necessary to prevent the market commodifying and controlling all of nature, and we also have a sense of what material goals we want to achieve, how then is it possible to organize production and consumption without a market? We draw on a range of influences, including Soviet cybernetics and mathematics, Chile's 'Cybersyn' programme, meteorology, and cutting-edge integrated assessment models (IAMs) used by climate scientists today. We have even constructed a Half-Earth socialist planning game to illustrate the difficult trade-offs such a society would have to navigate. We invite you to use the model online at http://half. earth, so that you, too, can be a Half-Earth socialist planner. With this preliminary model in hand, we try to envision what a complete global simulation might look like, and thus bring us closer to planning the world.

For dessert we indulge in utopian socialist fiction. Our story is a modest contribution to a literature that includes Thomas More's *Utopia* (1516) – which Karl Kautsky regarded as 'the foregleam of Modern Socialism' – Bellamy's *Looking Backward* (1888), William Morris' *News from Nowhere* (1890), Ursula K. Le Guin's *The Dispossessed* (1974), and many others. Instead of thinking abstractly about epistemology or on a global scale in terms of climate models, we focus our perspective on a neighbourhood and try to imagine quotidian life under Half-Earth socialism. What would it be like to work without the threat of unemployment? How would economic co-ordination function without money or a market? What might it be like to live in a world where nature can recover because half the planet has been rewilded? What does it mean to live in a society where the economy is consciously and democratically controlled?

Half-Earth Socialism's four-course structure offers food for thought in two ways. The first is as a prospectus outlining what would be necessary to transcend the environmental crisis. Some of our readers might be sceptical of our programme of rewilding and central planning, and therefore we invite them to use the book a second way, as a guide to utopian thought experiments. Proceeding along the same three levels of analysis, another budding futurist could make judgements that differ from ours. Perhaps the key philosophical principle of eco-socialism is not the unknowability of nature but something else, say, the hybridity of nature and culture. At the next step, our utopian reader could opt for nuclear power or even SRM, rather than a fully renewable system. Another choice could be creating a Two-Thirds Earth to reduce extinction to even lower levels ($0.67^{0.25} = 0.90$). Lastly, one might come up with a mode of economic distribution different from our cybernetic central planning and instead espouse an 'ecosystem of markets'.[67] Such a reader would devise a utopia different from Half-Earth socialism, and that is fine, for we do not pretend to have all the answers to the world's most difficult questions.

For too long the Left has been better at critique than creating its own positive proposals. In the rare chance that they take power, socialists will falter and fall without a programme to guide the transition beyond capitalism. Half-Earth socialism, we hope, is a vision of the future that can develop into a total alternative to capitalism, including everything from a plan for resource allocation to an outline of what life will *feel* like. We invite everyone from all liberatory traditions to join us in the revolutionary kitchen to think up many new recipes and work together to realize them. Indeed, we need many speculative contributions on the political horizon before it is suffused with a sulphurous mist and the future becomes as dim as the fixed grey skies of neoliberal hegemony.

1

Binding Prometheus

From the love of splendour, from the indulgences of luxury, and
from his fondness for amusement [man] has familiarised himself
with a great number of animals, which may not originally have
been intended for his associates.

–Edward Jenner

In the Sonoran Desert there are ziggurats of glass and steel, a
gleaming facility that looks like a Martian space station. Biosphere 2, as it is called, lies just beyond the exurban edge of
Tucson in the hamlet of Oracle, but once it seemed to be the
centre of the world. It was designed as a closed system able
to produce its own atmosphere, plant life, and water cycle by
mimicking the delicate balance of Biosphere 1 (i.e., Earth).
This monument to pharaonic ecology was built in 1989 by
members of a commune established twenty years before – the
Synergia Ranch – where everyone was to some extent a thes-
pian, cybernetician, gardener, sailor, and entrepreneur. John
Allen, the commune's leader, was the driving force behind Biosphere 2, but it was Ed Bass, another denizen of the Synergia
Ranch and scion of a Texan oil dynasty, who bankrolled it.
Bass and Allen, who each exhibited a hippie's heart and a
businessman's brain, hoped to understand the natural world
while earning a tidy profit. In the future, they believed, either
the environmental crisis would displace humanity to the stars
or a nuclear war would drive the species underground. Their
firm, Space Biospheres Ventures, was intended to cater to both
markets.

Biosphere 2 belonged to a long line of Cold War environmental research focused on escaping Earth's surface. Ecological concepts such as 'carrying capacity' first emerged in the 1950s research programme on 'cabin ecology': the study of human life in spacecraft, submarines, and bomb shelters.[1] CO_2 'scrubbing' technology, which underpins 'carbon capture and sequestration' systems today, was originally developed for the USS *Nautilus*, the first nuclear-powered submarine. The *Nautilus* could stay submerged longer than conventional submarines, which required advanced air-recycling technology (especially since the crew smoked).[2] The crucial moment during the ill-starred Apollo 13 mission was when the astronauts had to jerry-rig a CO_2 scrubber so they could survive the return voyage on their broken craft. Inspired by such artificial life-support systems, the military designer Buckminster Fuller coined the term 'spaceship Earth' in the early 1960s.[3] Fuller relied on military aircraft construction techniques for his designs, including the geodesic dome, that quintessentially 'eco' structure.[4] In 1962, almost thirty years before Biosphere 2 was built, the Ecological Society of America held a meeting to discuss the possibility of building a lunar base complete with a 'general life support system' able to circulate nutrients, oxygen, and CO_2 among cohabiting plants, animals, and astronauts.[5]

If the early Cold War sparked interest in closed systems, the 'New Cold War' of the 1980s revived this otherworldly market. Bass was confident that his verdant and ecological designs were superior to NASA's lifeless 'space cans' and sought to become a contractor for the new American space station *Freedom* before its planned launch in 1992.[6] Until the new eco-space industry took off, Bass hoped to recoup some of his investment by attracting tourists to his 'ecological Disneyland' in Oracle.[7] Within the gleaming walls of Biosphere 2, one could find five ecosystems reproduced in miniature: tropical rainforest, coastal fog desert, mangrove wetland, savannah, and ocean (with a coral reef!). There was also an agricultural area for crops and livestock, a laboratory, and living quarters for the crew. An assortment of mechanical and biological systems was needed to maintain Biosphere 2. The 'Technosphere' controlled air temperature and humidity, the 'Energy Center' ventilated air and regulated water temperature, and the massive mechanical 'Lungs' managed air pressure. There were also algal filtration to clean the water and a soil-bed reactor to aerate the soil and nourish its essential microbial ecosystem. Such systems were expensive, and Biosphere 2 – essentially a fancy greenhouse on a one-hectare plot – lost Bass $200 million. Given such immense resources, its purpose seemed modest enough: to keep eight 'biospherians' alive for two years without allowing anything (not even air) in or out of the complex. Two 'missions' were carried out at Biosphere 2 between 1991 and 1994. Both ended in disaster.

Before the end of the first mission, the biospherians had inadvertently managed to replicate many of the different facets of the environmental crisis in miniature – an appropriate outcome for an experiment conducted in a place named Oracle. Biosphere 2 was wracked by elevated CO_2 levels, extinction, loss of pollinators, dying coral reefs, invasive species, and eutrophication. Over-vigorous soil microbes, rather than fossil-fuel emissions, destabilized Biosphere 2's atmospheric chemistry: microbial breath reacted with concrete to create calcium carbonate, sequestering oxygen in the walls to the detriment of the living creatures within them. While the biospherians grew listless from the lack of oxygen, many creatures simply perished. Soon, all the pollinators were dead, leaving many plants to live on borrowed time.[8] The biospherians had to pollinate their crops by hand, an experience that eerily foreshadowed bee-less farming in Biosphere 1 two decades later.[9] Harvests were hard-earned but unsatisfying. Meat was a rare indulgence, and often there was little food of any kind. Starving crew members ate peanuts, shells and all, while the artificial biosphere around them withered into an empty husk.[10] Mission command finally relented and pumped a tonne of liquid oxygen into the complex. Although the biospherians danced with joy at this deus ex machina, the intervention nixed the project's scientific value as a study of closed systems.

Manna from heaven might have revived the biospherians, but it could not bring Biosphere 2 back to life. Some 19 of its 25 vertebrates went extinct, along with a majority of the 125 insect species.[11] The mass death of insects is a rare phenomenon, having occurred only during the third mass extinction (the 'end-Permian', some 252 million years ago) and the ongoing sixth. By contrast, invasive species thrived. A few stowaway ants and cockroaches, brought in via construction materials or soil, multiplied into rampaging swarms, including the 'crazy ant' (*Paratrechina longicornis*), which soon replaced all eleven ant species that had been carefully selected to cycle nutrients and disperse seeds.[12] Unchecked algal growth strangled the ocean, so that the world's largest captive coral reef had to be cleaned by hand. Bass and Allen's walled Eden had become a small, dying world.

The most important lesson salvaged from the wreckage of Biosphere 2 is the impossibility of controlling ecological systems even of a modest size. Shortly after the second mission, scientists argued in *Science* that the experiment made clear 'no one yet knows how to engineer systems that provide humans with the life-supporting services that natural ecosystems produce for free ... Despite its mysteries and hazards, Earth remains the only known home that can sustain life.'[13] Yet it seems that few have learned from Bass and Allen's folly. Neoliberals and their fellow travellers on the political centre and Left keenly support geoengineering as a solution to the climate crisis, as if the Earth system could be controlled like a thermostat. The 'Sixth Extinction', which will likely lead to the extermination of half of all animal and plant species by the end of the century, seems to bother few outside of the tiny conservation movement.[14] Yet according to leading ecologists, it is arguably '*the* most serious environmental problem', because of its irreversibility and impact on hyper-complex ecological systems.[15] Furthermore, the experience of Biosphere 2 discredits the popular approach of pricing 'ecosystem services', which have been valued at $125 trillion globally.[16] Instead of Oscar Wilde's cynic, it seems that it is now the naive environmentalist who 'knows the price of everything, and the value of nothing'.[17] If $200 million couldn't keep eight people alive in an ecosystem the size of two football pitches, how much is the world worth?

In this chapter we attempt to establish a system of thought fit for an age of environmental catastrophe. We begin by asking: What can we know? To what degree do we understand nature? What are the consequences of our ignorance, both to ourselves and to the natural world? These questions are similar to those posed by Friedrich Hayek about the economy in the 1930s and 1940s, in his effort to re-found liberalism as neoliberalism. By arguing that humanity could never comprehend the market, he endowed the economic sphere with ecological attributes – it appeared as a natural self-organizing system. We follow his example of digging deep into the epistemic layers of our beliefs, but whereas Hayek covered the market with the veil of ignorance, we drape that veil over nature. But what are the social, economic, and political implications of the impossibility of fully knowing nature? To answer this question, we take leave of 1990s Arizona and travel across the Atlantic Ocean and back two centuries to trace the origins of modern environmental thought.

The Battle of Dorking

Most of the epistemological frameworks for understanding the present environmental crisis can be traced back to one of three works written in 1798. Given the tumult of that era, this was not a particularly conspicuous year. Napoleon Bonaparte was busy fighting a fruitless campaign in Egypt and still a year from his coup d'état; the Haitian Revolution was only halfway through its thirteen-year duration; and it wasn't until 1800 that Alessandro Volta invented the battery. Yet, tucked between the folds of these better-known events, G. W. F. Hegel, Thomas Malthus, and Edward Jenner wrote texts that would in time come to define the three main environmental paradigms. Hegel's 'The Spirit of Christianity and Its Fate' was a personal essay in which he reflected on his recent reading in theology and political economy. Malthus' *Essay on the Principle of Population*, the most famous text of the three, has enjoyed a lasting influence on economics, demography, and population ecology. By contrast, Jenner had to self-publish *An Inquiry into the Causes and Effects of the* Variolæ vaccinæ because no one else would disseminate his findings from experiments on the smallpox vaccine. We see these three texts as belonging to an unspoken debate over the knowability of nature. Each of these thinkers is a primogenitor of one of the three great lineages in environmental thought: Hegel's Prometheanism, Malthusianism, and Jennerite ecological scepticism.

Before these men shaped the lineaments of environmental thought, they were swayed by the storm that was the French Revolution. Hegel, who is largely remembered as the grand old man of the conservative Prussian academy, was a radical student in his youth. One can find '*Vive la liberté!!*' scrawled in his yearbook from the University at Tübingen.[18] Meanwhile, Malthus' *Essay* immediately elevated him from obscurity as a parson in Dorking to fame and infamy, as well as giving him the first chair in political economy (at a college run by the East India Company). Despite his radical upbringing – his father had greatly admired Rousseau – Malthus wrote his *Essay* with the conservative hope of deflecting the 'blazing comet' of the French Revolution before it 'destroy[ed] the shrinking inhabitants of the earth'.[19] Jenner's pamphlet on his smallpox vaccine gained traction in 1799, when the French Wars sparked an outbreak of the 'speckled monster' in Britain.[20]

'The Spirit of Christianity and Its Fate' is one of Hegel's lesser-known works, but it is vital for our purposes. This is because it introduces his as yet unnamed concept of 'the humanization of nature' (*die Humanisierung der Natur*) for the first time – an idea that would animate much of his philosophical oeuvre, from *The Phenomenology of Spirit* to *Lectures on Aesthetics*.[21] The humanization of nature is the process by which humanity overcomes its alienation from nature by instilling the latter with human consciousness through the process of labour – transforming wilderness into a garden. Human labour is nature acting upon itself so it can become self-conscious, or, as Hegel puts it in the *Encyclopedia*, 'The goal of Nature is to destroy itself and to break through its husk of immediate, sensuous existence, to consume itself like the phoenix in order to come forth from this externality rejuvenated as spirit.'[22] On this basis, Hegel and his heirs fostered the belief that the domination of nature was both feasible and historically necessary.

In 'The Spirit of Christianity and Its Fate', Hegel grounds this concept in the biblical past. Between the Fall and the Flood, an era of which only 'a few dim traces have been preserved to us', humans struggled to 'revert from barbarism' and return to a state of grace by restoring their 'unity' with nature.[23] Nature rebuffed such efforts with the Flood, leaving humanity with three choices, each represented by a biblical figure.[24] Noah rebuilt the world through divine law, which controlled violence among people as well as violence between humanity and nature.[25] Rejecting this juridical peace, Nimrod subdued both people and nature to his will: 'He defended himself against water by walls; he was a hunter and a king.'[26] Abraham rejected Noah and Nimrod's social contracts because he aspired instead to be independent of nature and society.

By avoiding settlements and opting for herding over farming, Abraham was 'a stranger on earth, a stranger to the soil and to men alike'.[27] While Hegel originally wrote this essay to explain the rise of Jewish monotheism, he also devised his concept of the humanization of nature in its nascent form by contrasting the approaches of these three men. Noah subdued both humanity and nature to God's will, while Nimrod offered no reconciliation, only pure hostility. Abraham, too, represented a dead end by seeking merely independence, not freedom. True history, Hegel would come to argue, could only start once the opposition between nature and humanity had been reconciled through labour, that is, through the redirection of nature to human ends.[28]

The second ur-text from 1798, Malthus' *Essay on the Principle of Population*, was written to attack William Godwin, a proto-anarchist utopian who rose to prominence in 1793 following the publication of *Enquiry Concerning Political Justice and Its Influence on Morals and Happiness*. Godwin, who railed against the oppressive institutions of marriage and the monarchy, believed in humanity's ultimate 'perfectibility'.[29] His optimistic visions of 'cultivated equality' in a society where 'each man's share of labour would be light, and his portion of leisure would be ample' prompted Malthus to take up his pen to poke holes in Godwin's argument.[30] Of all the 'unconquerable difficulties' obstructing Godwin's utopia, Malthus focused on overpopulation. 'Population, when unchecked, increases in a geometrical [i.e., exponential] ratio', he asserted, while 'subsistence increases only in an arithmetical [i.e., linear] ratio.' In other words, population can increase exponentially – doubling every twenty-five years, he estimated – but agricultural yields increase by only the same amount each year. As Malthus wryly noted, 'a slight acquaintance with numbers will shew the immensity of the first power in comparison of the second.' The only way to return society to balance with its agricultural basis was through the 'checks' of famine, war, and disease.

The effects of such Procrustean demography 'must fall some where; and must necessarily be severely felt by a large portion of mankind'.[31]

Our third consequential text from 1798, Jenner's *Inquiry into the Causes and Effects of the* Variolæ vaccinæ, outlined a philosophy of nature that has rarely been acknowledged. It is better known for reporting on the 'challenge trials' Jenner carried out to test his smallpox vaccine. He had gleaned the idea that cowpox could inoculate against smallpox from the folkloric immunity of milkmaids to the disease. Jenner's vaccine was predated by the widespread but riskier practice of 'variolation', which had existed in much of Asia and West Africa before being brought to England by Mary Wortley Montagu in 1721. Nor was Jenner the first to intentionally use cowpox as a prophylactic, as a little-known farmer, George Jesty, had done so to protect his family in the 1770s.[32] Jenner's contribution lay in his scientific rigour and historical interpretation. His experiments created a more certain – but still far from complete – understanding of cowpox inoculation, which could be trusted, replicated, and disseminated.[33] This required Jenner to overcome several technical hurdles, such as differentiating bovine sores to make sure he was collecting cowpox virus for his vaccine instead of another pathogen, and preventing the vaccine from being contaminated with smallpox.[34]

What is more relevant for us is Jenner's explanation of why smallpox and its ilk existed at all. He argued that diseases were the result of humanity's unnatural domination of animals: 'The deviation of Man from the state in which he was originally placed by Nature seems to have proved to him a prolific source of Diseases.'[35] He was the first to posit that the attempt to control nature allowed new illnesses to emerge. Jenner saw that any animal used for any purpose might become a vector for disease:

The Wolf, disarmed of ferocity, is now pillowed in the lady's lap. The Cat, the little Tyger of our island, whose natural home is the forest, is equally domesticated and caressed. The Cow, the Hog, the Sheep, and the Horse, are all, for a variety of purposes, brought under his care and dominion.[36]

His emphasis on the link between animal and human health is reflected in 'vaccine', a term he coined from the Latin word for 'cow' (*vacca*). Remarkably, Jenner's historical insight has received little recognition. It is instead Rudolf Virchow, a German doctor three generations Jenner's junior, who is credited as the founder of 'one health' – the medical paradigm that encompasses human and animal well-being.[37] While it is easy to praise Jenner's skill as an experimenter, few have been interested in his philosophy of nature.[38]

These three texts from 1798 represent discrete epistemologies based on what can be known and controlled: nature, demography, or the economy. While Hegel did not share contemporary British political economists' enthusiasm for the market's self-correcting qualities, he conceded that its autonomy was necessary in a modern society.[39] Nature, on the other hand, would eventually be fully known and controlled by humanity. By contrast, Malthus believed that the biological life of humanity could be understood as easily as any other species because the law of population 'pervades all animated nature'.[40] The best instrument to manage demography would be the unfettered market, which is why he blamed the Poor Laws for spurring population growth and deepening poverty.[41] While Jenner did not directly write about the economy or demography, the implication of his argument was that diseases would cross species as long as humans exploited other animals. Let us now examine the afterlife of these ideas, from the eighteenth century to our times.

From 1798 to 2022

Hegel's most consequential heir has been Karl Marx. The seer from Trier incorporated much from the elder philosopher whole cloth into his socialist philosophy of history – including the humanization of nature.[42] As with Hegel and his 1798 essay, Marx wrote the 1844 *Economic and Philosophic Manuscripts* not for publication but to work through his own thoughts on a new subject: Hegel himself. In a revealing passage, Marx wrote that for man, 'in the working-up of the objective world ... nature appears as *his* work and his reality ... for he duplicates himself not only, as in consciousness, intellectually, but also actively, in reality, and therefore he contemplates himself in a world he has created.'[43] Marx foresaw that eventually, at the end of history, 'the objective world becomes everywhere for man in society the world of man's essential powers ... all *objects* become for him the *objectification of himself*, become objects which confirm and realize his individuality'.[44] Humanity would no longer regard nature as an alien, threatening force but a reflection of human consciousness after natural forces have been redirected towards human ends.

Marx has a reputation as a 'Promethean', that is, he is seen as a thinker who believes that the total control of nature is necessary for human freedom. Prometheus, described by Hesiod as the Titan 'of the intricate and twisting mind', stole fire from Zeus by secreting it in 'the hollow of a fennel stalk' so that he could give it to a benighted humanity.[45] Zeus punished Prometheus by leaving him impaled on a mountain in the Caucasus where an eagle ate his regenerating liver every day. It seems that the Promethean question can be distilled to one's relationship to birds. While Jenner wrote a remarkable ethology on the cuckoo – this research, not his vaccination experiments, earned him membership in the Royal Society – Marx mocked those who 'childishly wonder at the cuckoo laying eggs in another bird's nest'.[46] Unfortunately, Marx's

disdain for avian life was echoed by twentieth-century socialists. Leon Trotsky hunted ducks. Stalin killed his pet parrot
with his bare hands. One of the few exceptions is Rosa Luxemburg, who used to identify species she saw and heard from
her prison cell during World War I.[47]

Marx's Prometheanism stemmed not from a disagreement
with Jenner (whom he never mentioned), but from his rancour
towards the anti-utopianism of Malthus, 'that master in plagia-
rism' and 'professional sycophant of the landed aristocracy',
who faced more vitriol than any other 'bourgeois economist'
criticized by Marx.[48] This was because Marx was well aware
that even his lightly sketched utopia of the communist 'realm
of freedom' could be attacked on Malthusian grounds.[49] In
his influential reconstruction of Marx's conception of communism, philosopher Bertell Ollman argued that the new joyful,
free, and stateless society depended on unprecedented abundance wrested from nature. After all, if there is no scarcity,
work would be merely for pleasure and there would be no
reason to steal.[50] The standard socialist response to Malthus
seems to have changed little since Friedrich Engels claimed in
1844 that 'science increases at least as much as population'.[51]

Although recent exegesis suggests that Marx blunted his
harsh Promethean edge in old age, it is nevertheless the Promethean Marx who overshadows the socialist tradition.[52]
Trotsky believed that a communist society could 'cut down
mountains and move them ... and repeatedly make improvements in nature' so that eventually '[man] will have rebuilt
the earth ... according to his own taste'.[53] While the USSR
boasted ecological thinkers of the first order, such as Vladimir
Vernadsky and Vladimir Stanchinsky, and innovated a system
of protected wilderness areas (*zapovedniki*), it also produced
many Promethean schemes both imagined and realized.[54]
In 1959, Boris Lyubimov proposed warming up Russia by
damming the Bering Strait, which would destroy the Arctic's
ice and affect the region's currents ('the central heating system

of our earth'). 'Man's ability to triumph over nature is limitless,' Lyubimov proclaimed, 'modern science and technology have armed him with powerful means of reclaiming barren deserts, subjugating the cosmos, changing the climate, and mastering the mighty and inexhaustible energy of the atom.'[55] Similar to Lyubimov's search for Earth's optimal temperature was Mikhail Budyko's plan in 1974 to inject 200,000 tonnes of sulphur into the stratosphere to reverse fossil-fuel-induced heating.[56] (Yes, a Soviet climatologist was the first to advocate 'solar radiation management'.) Lyubimov's and Budyko's schemes were fortunately not carried out, but actually existing Prometheanism was still a disaster. To cite but one example, the world's fourth-largest lake, the Aral Sea, was destroyed by Nikita Khrushchev's 'virgin lands' campaign in the late 1950s, which diverted the Syr and Amu rivers to irrigate fields of cotton and wheat in central Asia.[57] By the 1980s the lake had shrunk by half as the rivers dried up, and today a mere tenth remains. For years now, the dying giant has shed 'dry tears' of pesticides and salt, wind-scoured from the dry lakebed, raining death on farmers and former fishers.[58]

Our point here is that Marxism cannot simply be greened by reading *Capital* with viridian-tinted glasses, as some theorists have tried.[59] Prometheanism is so ingrained in Marxist thought that it must be confronted, refuted, and extirpated so that socialism can be made fit for an age of environmental catastrophe. Alas, there is little sign such reckoning is taking place. Indeed, much of the Left appears to be becoming *more* Promethean as the biosphere dies. In 2013, 'accelerationists' Alex Williams and Nick Srnicek called for a 'Promethean politics of maximal mastery over society and its environment'.[60] Four years later, *Jacobin* published a special issue on the environment which included essays praising geoengineering and nuclear power to undergird cornucopian communism.[61] Holly Jean Buck, a socialist and self-described 'geoengineer', warns the Left that entrepreneurs who are 'taking action, being creative, or disrupting' are 'the wrong focus of critique'.[62] She speculates that one day there might be carbon-credit 'gift cards' and an AI-controlled SRM (a true Skynet).[63] Marxists – those hard-nosed critics of capitalism – still see the domination of nature through a mist of schwärmerei rather than the poisonous dust of the Aral Sea.

Malthusianism has aged just as poorly as Prometheanism in the centuries since 1798. Although Malthus has never fallen out of favour among economists, ecologists, and demographers, his influence within the environmental movement crested during the so-called 'Malthusian moment' of the 1960s and 1970s.[64] Part of that shift was due to Paul Ehrlich's *The Population Bomb* (1968). The young lepidopterologist had become obsessed with the problem of population after a vacation in India, where he was horrified to encounter 'people eating, people washing, people sleeping. People visiting, arguing, and screaming. People thrusting their hands through the taxi window, begging. People defecating and urinating. People clinging to buses. People herding animals. People, people, people, people.'[65] Infamously, Ehrlich's book begins with the prediction that 'the battle to feed all of humanity is over ... in the 1970s and 1980s hundreds of millions of people will starve to death in spite of any crash programs embarked on now.'[66] Doom sells well – with some 2 million books sold and twenty invitations to Johnny Carson's *Tonight Show*, Ehrlich soon became a household name.

The same year *The Population Bomb* came out, population biologist and strident white nationalist Garrett Hardin published his short article 'The Tragedy of the Commons'. He used the metaphor of a pasture overgrazed by selfish peasants to warn how certain races, religions, or classes would exploit the largesse of the welfare state to subsidize the 'freedom to breed'.[67] In 1974, he outlined his 'lifeboat ethics' as a Malthusian analogue to Fuller's 'spaceship Earth'. 'Metaphorically, each rich nation amounts to a lifeboat full of comparatively rich people', Hardin explained, while the poor nations are life-boats so crowded that 'the poor fall out' and hope to benefit from the 'goodies' in the rich lifeboats.[68] The Malthusian imperative was simple: let them drown.

There are different variants of Malthusianism, but often little separates the genocidal from the respectable strains. This split can be found in the example of Malthus himself, who condoned starvation for poor Britons, but was appalled by the genocide of Indigenous nations in North America – a tragedy that he blamed on American population growth.[69] It is ironic then that Hardin warned of a 'passive genocide' of whites caused by people of colour 'outbreed[ing]' them, and called for the US state to coercively control the reproduction of non-white women.[70] Despite his abhorrent rhetoric, Hardin remains one of the most cited scholars of all time and his framework is often taught *sans critique* to undergraduates.[71] Many popular environmentalists, such as David Attenborough and Jane Goodall, share the Malthusian view that the human population should be much smaller, though they do not openly support coercive measures.[72] One of the more moderate Malthusian proposals is a cap-and-trade programme for pregnancy permits, while others have flirted with forced abortions and sterilization.[73] The appearance of an 'eco-fascist' mass shooter represents the logical, albeit extreme, terminus of Malthusian thought.[74]

Jenner's position of ecological scepticism was the least developed of the three paradigms examined here, but we can fruitfully extrapolate from its basic principles. Rather than seeing disease as a timeless burden, Jenner historicized it as an interspecies relationship. It is now widely accepted that prior to the rise of animal husbandry, humans suffered no disease apart from the occasional parasite or unlucky infection, suggesting that the pathogens that now plague humanity ultimately come from other animals.[75] In the context of the 300,000 years that human beings have existed, zoonoses only emerged relatively recently, following early animal domestications 10,000 years ago. Measles likely evolved from rinderpest (a bovine disease) 7,000 years ago; influenza began infecting humans 4,500 years ago after waterfowl were domesticated; leprosy came from water buffalo; and variants of the common cold can be traced to horses.[76] Jenner's own specialty, smallpox, probably originated 4,000 years ago in eastern Africa, when a gerbil virus jumped to the newly domesticated camel and thence to humans.[77] A revealing contrast can be made with Indigenous nations in the New World, where even large urbanized societies such as the Inca or Aztecs suffered little disease before 1492 because few animal species had been domesticated.[78] While Hegel did not consider Abraham's pastoralism as part of the humanization of nature, Jenner recognized animal husbandry as one of the most consequential and dangerous ways humans shape life on Earth.

Jenner's ecological scepticism can be applied more broadly across the span of the environmental crisis because zoonotic diseases can be engendered or spread by practically any ecological disturbance: animal testing (Marburg virus), cattle ranching (Junin virus), deforestation (malaria), factory farms (MRSA), factory farms *and* deforestation (Nipah virus), pets (psittacosis), biodiversity loss (West Nile virus), dams (schistosomiasis), habitat fragmentation (Lyme), and the exotic-animal trade (SARS). Not only has the scale of humanity's attempted control of nature never been greater, but its pace of growth has become ever faster. If a first epoch of disease lasted from 10,000 years ago to the mid-nineteenth century as new diseases accompanied the expansion of animal husbandry, then a second, much shorter period of effective public health, vaccines, and antibiotics ended a century later in the 1970s. Epidemiologists warn of a 'catastrophic storm of microbial threats' in the ongoing third era of disease that could return us to the epidemiological Stone Age.[79] Ebola's emergence in 1975 is a useful marker of the transition from the second to the third period. In the aftermath of that outbreak, it was already apparent to contemporary scientists that 'the larger the scale of man-made environmental changes and the more they involve areas little frequented by man, the greater must be the probability of emergence of a zoonosis'.[80]

The Jennerite solution to the problem of disease is to undo much of the humanization of nature and leave it forever incomplete. This is less a philosophical nicety than a matter of public health. After the 2003 SARS epidemic, the *American Journal of Public Health* published an opinion piece demanding a change in how 'humans treat animals – most basically, ceasing to eat them or, at the very least, radically limiting the quantity of them that are eaten' as a basic measure of disease prevention.[81] The American Public Health Association has long called for a moratorium on factory farms.[82] Wildlife conservation is now seen as an essential component of public health, because nature preserves act as cordons sanitaires.[83]

It is unclear if Jenner realized that zoonotic genesis was not only an ancient phenomenon but also one occurring in his own lifetime. Take cholera, for example. The lush coastal landscape of the Bay of Bengal had for millennia sustained a forest of extraordinary biodiversity, including the bacterium *Vibrio cholerae*, which preyed on crustaceans. In the decades that followed the East India Company's conquest of Bengal in the 1760s, 90 per cent of the region's mangroves were cut down and replaced with embankments and rice plantations. To survive this ecological disturbance, the *Vibrio cholerae* evolved 'a long, hairlike filament at its tail that improved its ability to bond to other vibrio cells ... form[ing] tough microcolonies that could stick to the lining of the human gut'.[84] In 1817, heavy rainfall flooded the town of Jessore with cholera-infested waters from the forest, sparking the first cholera epidemic. The terrifying new illness weakened its victims with diarrhoea and vomiting, turned their skin blue, and could deliver a swift death within a matter of hours.[85] From Jessore the dread disease marched inexorably across Eurasia and reached Berlin in 1831. Hegel fled with his family to the suburbs for safety but could not escape cholera's grasp – a poetic fate for the prophet of the humanization of nature.[86]

Though it is often forgotten by Prometheans, the myth of their eponymous hero was not just a paean to the human conquest of nature but also a warning against such hubris. This is clear if one reads Hesiod, whose mythology resonates both with Hegel's early essay and present-day epidemiology. The period before Prometheus' theft sounds remarkably Edenic, where 'the races of men had been living on earth free from all evils, free from laborious work, and free from all wearing sicknesses'.[87] This pre-civilizational good life is what Marshall Sahlins once called the 'original affluent society'.[88] After Prometheus stole fire from the gods, they sent the beautiful Pandora, who opened her jar and released 'sad troubles for mankind'.[89] After the Promethean revolution, the ancient Greeks confronted nature much like Hegel's Israelites, for the land and sea became 'full of evil things' and 'there are sicknesses that come to men by day ... bringing sorrow to mortals'.[90]

The Automatic Subject

Capitalism marks a break in human history because the pace and direction of progress – the humanization of nature – have been usurped by capital. Unlike in previous historical eras, we now live in a society driven by the *capitalization* of nature, because humans no longer consciously direct this process. In this way, Marxists and neoliberals share a strikingly similar understanding of the market as an unconscious, all-powerful force, the difference being that the former abhor it, while it is worshipped by the latter. This approach is hardly coincidental because, as we shall see, the neoliberal conception of the market is derived *from* Marxism. However, before we delve into this snarled intellectual history, we must first understand what capital is.

For Marx, capital is less a thing than a set of social relations. It emerges from the relationship between capitalists and workers, after the latter have been separated from the 'means of production' (e.g., farmland, tools) and forced to sell their labour on the market, and it is sustained by capitalists competing against each other in the pursuit of maximal profit. Thus, unemployment and the need to make the going rate of profit constrain the freedom of both classes, albeit in unequal fashion. Capitalism is an unusual social form in that the elite do not directly control the labour of the lower classes, as was the case with pre-capitalist societies (e.g., via the corvée or tithe), nor are investment decisions guided by a product's direct utility ('use-value') but by the opportunity to garner profit from selling it to others ('exchange-value'). Despite being a product of human social relations, capital confronts said humans as an autonomous force guiding their actions: the market takes on a life of its own. Capitalism, then, is like Sardinian throat singing, in which four performers produce *la quintina* – the illusory fifth voice.

To reset this problem in Hegelian terms, pre-capitalist history was propelled by the relationship between 'master' and 'slave'. While the former risks death in the fight for recognition, the latter submits to survive. The slave's labour transforms nature and humanizes it for the sake of the master, but through this very process, the slave learns and thus changes herself while the master stays the same. Counter-intuitively, the capitalist is not the 'master' in a capitalist society, as both Hegel and Marx realized.[91] The master is capital itself. Marx calls it the 'automatic subject', an unconscious force: 'constantly assuming the form in turn of money and commodities, it changes its own magnitude, throws off surplus-value from itself considered as original value, and thus valorizes itself independently.'[92]

Capital's self-expansion becomes an end to itself, regardless of the consequences to society and nature. In this 'inverted world', labour no longer produces an environment that reflects human consciousness, but 'the world of wealth expands and faces [the worker] as an alien world dominating him'.[93] In the factory, the worker does not use the means of production; rather, it is the 'means of production that make use of the worker'.[94] In his unpublished *Jenaer Realphilosophie* (which Marx never read), Hegel prefigures Marx's analysis to a surprising degree by characterizing the market as 'a monstrous system of community and mutual interdependence in a great people; a life of a dead body, that moves itself within itself, one which ebbs and flows in its motion blindly'.[95]

Compared with ancient tyrants or mediaeval kings, whose wills were constrained only by the wills of other masters and nature itself, capitalists make a strange elite. Pressures from the market dehumanize the capitalist, reducing her to the hybrid creature Marx called 'capital personified and endowed with consciousness and a will'.[96] To illustrate the dual imprisonment of both worker and capitalist, Theodor Adorno and Max Horkheimer turned to *The Odyssey* for inspiration. They compare the rowers of Odysseus' ship, who avoid the Sirens' treacherous songs by plugging their ears with wax, to factory workers whose craft has been stripped of all beauty and pleasure. Rather than enjoying their labour, the rowers must 'look ahead with alert concentration and ignore anything which lies to one side'.[97] Odysseus, the fable's capitalist, is also unfree. He is strapped to the mast so that he does not surrender to the Sirens, those representatives of life's sensuous and aesthetic pleasures. There might be less sweat on his brow, but otherwise Odysseus' fate is no freer than his men's. For Adorno and Horkheimer, the parable represents the moment when 'the enjoyment of art and manual work diverge as the primeval world is left behind' because of capital's 'inescapable compulsion toward the social control of nature'.[98]

If the helmsman of Spaceship Earth is capital – rather than the capitalist and certainly not the worker – then it becomes clear not only why we have entered the choppy waters of the environmental crisis but also why it seems impossible to change course.[99] After all, it has long been clear what must be done, such as switching to renewable energy, expanding nature preserves, and eating less meat. Within the dry technocratic tomes produced by the Intergovernmental Panel on Climate Change (IPCC), one can find surprisingly radical proposals reflecting this consensus. In addition to abjuring 'cost-benefit' analysis (marking a shift from money being the measure of all things), the IPCC has called for carbon-neutral building codes, a ban on new coal-fired power plants, and reducing cars through 'effective [urban] planning'. The IPCC even blames the belief in 'individual autonomy' and 'free-market ideology' for allowing climate denial to fester, as well as 'vested interests' and 'industry group lobbying' for blocking reform.[100] Despite our knowledge of what needs to be done, carbon emissions increase and mass extinctions continue relentlessly. Capital is at the helm, blindly steering the ship of fools towards ecological disaster. Unable to feel the wind or listen to its shouting passengers, capital can sense only price signals to guide its passage. In this way, capital destroys the world it cannot see.

If capitalism is a society characterized by unconscious control, then socialism must be the restoration of human consciousness as a historical force. In practice, this means that the market must be replaced by planning. While this might appear a natural corollary of Marxist analysis, there are actually few theorists who follow the critique of capital's 'inverted world' to its logical conclusion. An exception is Otto Neurath, a remarkable but largely forgotten polymath of early twentieth-century Vienna. The kernel of his philosophical system was the rejection of 'pseudorationality' – the belief that any single metric could guide all decisions.[101] While profit was his main target, he was equally appalled by other universal metrics such as

Karl Popper's 'falsifiability', utilitarian 'pleasure', or contemporary socialist proposals predicated on energy or labour time.[102] Neurath saw that even a socialist economy based on a universal equivalent would lack the necessary conscious control that could prevent irrational outcomes, a critique applicable to the influential 'market socialism' of Oskar Lange and Abba Lerner. As early as 1919, Neurath criticized the desire of fellow socialists to 'hold on to the split and uncontrollable monetary order and at the same time to want to socialize' as 'an inner contradiction'.[103]

Neurath came to these conclusions by studying ancient and contemporary examples of economies based on 'natural' (or *in natura*) units of discrete physical things rather than money. In 1906, he finished his doctoral dissertation on the non-monetary economy of ancient Egypt. He was convinced that money did not necessarily represent an advance in economic history, as 'the large store-keeping economy of the ancient Egyptian kings and princes, with their accounting facilities, their wages in kind and other institutions was on a much higher level than the Greek money economy of the fourth century [BCE].'[104] Neurath employed his insights from ancient Egyptian economics to understand the collapse of the monetary economy during the Balkan Wars (1912–13), a conflict he observed first hand. His subsequent study on in natura war economics secured him a position as a planner in the Austrian War Ministry during World War I. These experiences led him to see in natura calculation as the solution to the problem of pseudorationality. After all, Neurath argued, there were no 'war units' to guide a battleship commander's decisions. What mattered were incommensurate things: 'the course of the ship, the power of the engines, the range of the guns, the stores of ammunition, the torpedoes, and the food supplies'.[105] In an emergency, prices fail to convey any information at all.

Neurath reforged his theory of the sword into the ploughshare of socialism when he was appointed head of the Bavarian

Soviet Republic's Central Planning Office in 1919. At the time, Lujo Brentano (a fellow economist and member of the Soviet) dismissed Neurath as an 'ancient-Egyptian romantic economist', which, admittedly, was a pithy summary of his world-view.[106] Neurath outlined his theory of planning in his pamphlet 'Through War Economy to Economy in Kind' (1919). The war had usefully shown that profit could be disabled as the guiding metric for investment, a development that would be continued under socialism.[107] The socialist economy would have a 'central office for measurement in kind' (i.e., in natura) that would draw up several 'total plans' based on available resources and chosen after 'the direct consideration of various possibilities' by a 'decisive central body'.[108] In natura calculation would banish the 'veiling concealments' of money so that 'everything becomes transparent and controllable.'[109]

Neurath's pamphlet attracted the attention of Ludwig von Mises, his former colleague at the War Ministry and future co-founder of the Mont Pèlerin Society. Mises' critical 'Economic Calculation in the Socialist Commonwealth' (1920) not only sparked the 'socialist calculation debate' about the possibility of economic planning but also was the first text of the nascent neoliberal movement. It is revealing that in a period abounding with socialist talent, it was not Luxemburg or Lenin but the relatively unknown Neurath who was the object of Mises' ire.[110] This is likely because Neurath outlined the principles of socialist governance with rare clarity; Hayek later complimented Neurath's early pamphlet 'Through War Economy to Economy in Kind' as the 'most interesting' opposing contribution to the socialist calculation debate.[111]

Mises' defence of capitalism was constructed in the mirror image of Neurath's critique. Mises considered money's universality as the very basis of economic rationality.[112] While he conceded that consumer goods could be distributed by a Neurathian planned economy, firms making 'higher-order' goods (i.e., 'intermediate goods' used by producers rather than consumers, such as steel) faced limitless economic alternatives and thus needed the price system as 'a guide through the oppressive plenitude of economic potentialities'.[113] Markets decide whether the steel goes into wind turbines or luxury SUVs. Mises' argument may have been the opening salvo in the 'socialist calculation debate', but it was not the strongest. The market socialists Lange and Lerner dispatched Mises in the mid-1930s, when they showed how a planning board could act as an auctioneer to efficiently balance supply and demand. Yet to make this argument, they concurred with Mises' critique that a universal equivalent was necessary for rational calculation.[114] Neurath could not convincingly articulate how rational economic decisions were to be made using in natura calculation.[115] Less than a decade later, innovations in mathematics – especially linear programming – provided tools that would aid in natura planning, though Neurath remained unaware of these advances by the time of his death in 1945.[116] (We examine these innovations in detail in chapter three). Nonetheless, it would be Lange and Lerner's 'market socialism' that would become dominant in socialist thought, and Neurath's in natura planning would be largely forgotten. There would be, however, another twist in the socialist calculation debate.

As Mises stumbled, his protégé Hayek stepped in with two landmark essays that revolutionized neoliberal thought: 'Economics and Knowledge' (1937) and 'The Use of Knowledge in Society' (1945). While Hayek agreed with Mises that prices helped individuals simplify economic decisions, he took Mises' argument a step further by emphasizing that the crux of the problem was not simply mathematical but epistemological. Lange, Lerner, and other market socialists, Hayek argued, thought planning was a matter of finding the mathematical solution to this 'optimum problem' so that 'the marginal rates of substitution between any two commodities or factors must be the same in all their different uses.' Thus, the state-as-auctioneer could arrive at such a solution through trial-and-error until supply and demand is balanced (i.e., Walrasian tâtonnement). Yet Hayek pointed out that a planning board could only solve its calculations if all the necessary 'data' were 'given'. More realistically, Hayek thought that the information the planners needed would be 'dispersed bits of incomplete and frequently contradictory knowledge'.[117] Only by acting through markets could individuals contribute their small fragment of knowledge and therefore have 'their limited individual fields of vision sufficiently overlap so that through many intermediaries the relevant information is communicated to all.' In this way, the market was best understood as a 'mechanism for communicating information' that had the same effect as 'one single mind possessing all the information'.[118] Surprisingly, Hayek's conception of the market resonates significantly with Marx's – capital is an unconscious but powerful force conjured by social relations.

Hayek's epistemic critique of socialism is a powerful one the Left has yet to refute. To do so, it is necessary to follow Hayek on his preferred intellectual terrain – the study of ignorance, or 'agnotology'.[119] While neoclassical economists (market socialist or not) assume that equilibrium is the natural state of things and that it is fine to assume 'perfect knowledge' to make their models work, Marxists and neoliberals see the market as a chaotic and unknowable system. Markets are riddled with monopolies, gluts, inequality, and environmental 'externalities'. No one is in control and no one knows what is going on. Frank Knight, a proto-neoliberal theorist, thought that entrepreneurs 'cannot be well or truly informed regarding the markets' while firms were mere 'groups of ignorant and frail beings'.[120] Hayek thought that firms didn't even know things as basic as their own cost curves.[121] The neoliberal capitalist is less a cold, rational *Homo economicus* than a pin-striped Neanderthal.[122] Rather than refuting Neurath's critique of the market as a decentralized and irrational system, the neoliberals staked everything on making virtues of the market's weaknesses. Planning would be impossible, but so too would even the moderate reforms of centre-left Keynesian or neo-classical economists. If the market is so opaque, how could a government intervene to make up the short-fall in demand? How could a government know the true cost of environmental damage in order to set an externality tax? According to Hayekian logic, the only solution to any market failure is more markets.[123]

To naturalize the market, neoliberals described it with biological and even theological metaphors. Hayek, who came from a family of biologists, saw the market as a 'highly complicated organism'.[124] This approach contrasted with that of neoclassical economists, who used Newtonian physics as their point of reference, as if the economy were a billiards game whose outcome could be predicted. There was also a strong theological streak to Hayek's thought, as if the market were divine and economists its humble priests. If the workings of the market were akin to the 'movement of the heavenly bodies ... directed by forces which we did not know,' then economists should comport themselves like sixteenth-century Jesuits 'who emphasized that what they called *pretium mathematicum*, the mathematical price, depended on so many particular circumstances that it could never be known to man but was known only to God.'[125]

The environmental crisis, however, strains the neoliberals' foundational postulates because they have to decide what is more complex and unknowable – the market or nature? Judging by their treatment of the environmental crisis, the neoliberal view is clear: despite nature being the wellspring of their metaphors of the market, neoliberals believe that the market can be used to know and control nature. This is why most neoliberals have only contempt for science, a fallible way of knowing nature compared to the market. Some have claimed that the market could predict the state of fish populations better than marine biologists, while others believed that the market and not government regulators like the Food and Drug Administration should decide whether new medical treatments were safe.[126] The neoliberals' forked-tongue denial of climate science – they are not so stupid as to believe their own rhetoric – buys time for their preferred solution of SRM.[127]

Of their many different approaches, cap-and-trade is the neoliberals' most sophisticated and widespread environmental policy, one that has been applied to everything from dredging cold-water sponges, mercury pollution, acid rain, and carbon emissions. Cap-and-trade creates a fungible right to inflict environmental harm, and the market rather than scientific expertise or democratic opinion decides who exercises such rights. John Dales, who devised the cap-and-trade approach in 1968, was unusual for caring more about nature than most neoliberals (he was an avid birder), but even he believed that clean air and water were 'luxury goods' rather than necessities.[128] The neoliberals' insouciance towards the destruction of nature is rooted in their belief that the problem is essentially a matter of aesthetics, not ecology or health. In his exchange with Neurath, Mises acknowledged that price might not convey the true value of a waterfall when compared with the potential profit of building a dam, but he was only concerned with the 'beauty' of the waterfall.[129] In an era of climate change, this logic leads to geoengineering despite its manifest threat to the essential and extremely complex Earth system. Neoliberals willingly gamble with something as risky as SRM rather than countenance restrictions on their revered market.

To get a sense of how foolish it is to think that privatized geoengineering will produce an optimal climate, it is worth remembering the shock of the ozone hole. High up in the stratosphere, the ozone layer safeguards life by blocking the sun's dangerous ultraviolet radiation. The chemistry of ozone depletion had become an object of concern since 1970 thanks to Paul Crutzen – of 'Anthropocene' fame – who worried about the impact of nitrogen pollution from supersonic planes.[130]

Four years later, Mario Molina and F. Sherwood Rowland studied how chlorofluorocarbons (CFCs) posed an additional threat to the ozone layer.[131] While supersonic flight never quite took off, there was plenty of CFC pollution because the compound was a popular coolant (having replaced lethal gases in refrigerators).[132] In laboratory conditions, CFC molecules appeared inert, but up high in the stratosphere, exotic reactions involving iridescent polar clouds began to take place. Still, unaware of what was happening in the Antarctic sky, scientists believed it would take fifty to one hundred years before the ozone layer shrank even 5 per cent.[133]

Atmospheric conditions in the South Pole, however, differ from warmer climes. In 1982, Joseph Farman, a British geophysicist, took routine atmospheric measurements at his Antarctic research station in Halley Bay, as he had done every year since 1957. To his surprise, he found that ozone had dropped 40 per cent. He trawled through data from previous years and realized that the decline had already begun in 1977 without him noticing. To make sure it wasn't just a local anomaly, Farman led his team the following year on a 1,600-kilometre trek to take measurements on the other side of Antarctica but found the same result.[134] He published his findings in *Science* in 1985, by which point the South Pole's ozone had declined by half. The combination of extreme swings in seasonal daylight with the unusual chemistry of polar stratospheric clouds meant that the air above Antarctica had high concentrations of chlorine – radicals that could easily pry apart CFCs. The gravity of the situation was quickly recognized, and by 1987 governments from around the world had agreed to the Montreal Protocol, which banned the use of CFCs everywhere. It took time for this impressive diplomatic feat to have some effect, and the ozone layer continued to thin until the late 1990s, when it was a mere 33 per cent of its normal thickness.[135] Even now it is nowhere close to full recovery. Perhaps the strangest aspect of this story was that

NASA satellites had long taken measurements of Antarctic ozone *every day* and recorded the same decline as Farman, but the results were so anomalous that the data-processing programme junked them.[136]

The ozone crisis is not just a useful allegory to warn against an overweening geoengineering programme but is directly relevant because SRM might itself damage the ozone layer.[137] To overcome this problem, David Keith, a leading geoengineer, has suggested swapping sulphur aerosols for neutral calcite (a calcium carbonate compound like the one that robbed the biospherians of their oxygen). He even claims that this modified SRM 'may cool the planet while simultaneously repairing the ozone layer.'[138] Once again, it turns out the stratospheric chemistry isn't so simple. Daniel Cziczo, an atmospheric scientist, criticized Keith's proposal for overlooking how calcite would react to sulphates already present in the stratosphere to create compounds that 'effectively promote ozone loss'. He went on to argue that Keith had relied on an 'overly simplistic set of assumptions' and failed to identify the 'unintended consequences' of his proposal.[139] (Perhaps it is fitting that CFCs were invented in 1930 by Thomas Midgley Jr., a researcher at General Motors and a modern Pandora, whose other infamous innovation was leaded gasoline.)[140] Epistemic humility is a virtue when confronting a system as vast and complex as the climate, but this is a lesson that geoengineers and their neoliberal admirers refuse to learn.

La Végétation en Marche

As the term itself implies, the Sixth Extinction is not the first to be endured by Earthlings. The steady drum beat of species loss heard today echoes a catastrophe from 375 million years ago: the Late Devonian Extinction. The trigger for that ancient apocalypse was the plant kingdom's conquest of land. Apart from fungi and cyanobacteria, life had been confined to the seas, but newly evolved vascular plants could withstand the elemental rigours of earth and air. Their roots penetrated Earth's craggy surface and broke up rocks to create the first soil. Rain swept away newly exposed minerals into the seas, causing massive algae blooms, and soon oxygen-poor dead zones proliferated along the world's coasts. Aggravating this threat to marine life was hydrogen sulphide – toxic refuse left behind by tiny organisms feasting on the algae. Vegetal life ventured deeper into hitherto lifeless continents, creating an untrammelled leafy empire devoid of herbivores larger than insects. Plants not only greened the land in their image but remade the heavens too. Leaves, trunks, and roots captured atmospheric CO_2 and diminished the greenhouse effect. In this cooler world, inland glacial seas dotted the landscape, and coastlines advanced as sea levels lowered, imperilling yet more aquatic species. Like the humans of Biospheres 1 and 2, ancient plants wrecked their earthly home with the plagues of eutrophication, climate change, and mass extinction.[141]

One would think that a difference between humans and *Archaeopteris* – an extinct genus of fern-like trees – is that our species is aware of its impending doom. Yet, given our continued inaction in the face of disaster, this is difficult to discern. The only way to demonstrate our vaunted powers of consciousness is to end the thoughtless capitalization of nature and to limit our species' interchange with nature through extensive but careful planning. Otherwise, a world of ever greater inequality, disease, climatic disaster, and ecological impoverishment awaits. Reversing these trends requires a thorough understanding of the problem as well as a variety of potential solutions. To this end we have drawn on an eclectic collection of thinkers in this chapter to inspire our own philosophy of nature.

Many of these thinkers use nautical metaphors of some kind, but perhaps the one most fitting is Fuller's 'spaceship Earth', a concept that was refined by the economist and anti-nuclear activist Kenneth Boulding.[142] In an essay published in 1966, Boulding divided history into two periods: the 'cowboy economy', founded on the seemingly limitless resources of the frontier, and the 'spaceman economy', in which Earth 'has become a single spaceship, without unlimited reservoirs of anything, either for extraction or for pollution'.[143] Spaceship Earth has generally been understood in two ways. The first is that Earth could be fully known, scaled down, and made modular, so it could be contained in an ark or perhaps a whole armada of spaceships Earth. It is this approach that inspired Biosphere 2 in the 1990s and the parade of billionaires' space colonization fantasies in the 2020s.[144] The second interpretation is that there is only one vessel, Earth, that can contain and sustain both humanity and nature, and therefore we must be careful to properly maintain this vital craft. One can go beyond Boulding by giving his framework a Jennerite twist: Earth is a natural machine, both ancient and alien, whose operating systems we will never fathom, and therefore it is wisest to let the ghost in the shell control the circuitry even if we do not always understand it ourselves.

The tension between the natural philosophies of Hegel and Jenner is evident even in Prometheus' name, which means 'forethinker'. Prometheanism is predicated on the ability to intelligently intervene in nature, but such action is impossible on Hegel's own terms because knowledge is produced through action. Therefore, the humanization of nature must proceed in conditions of ignorance, with all the danger that entails. This is why repercussions, such as zoonoses, are the unpredictable results of such action, and thus we must limit the admixture of human consciousness and self-willed nature to protect both spheres of life. By contrast, neoliberals – the bastard heirs of Hegel – seek not the humanization of nature but its complete capitalization. They are strange Prometheans to be sure. Rather than the conscious control of nature, the neoliberals seek one unconscious realm (nature) to be subdued by another (capital). It is revealing that the attendees at the first Mont Pèlerin Society meeting nearly named their organization after Prometheus.[145]

More than twenty years ago, geoengineer David Keith imagined a future where the world had become a human 'artifact'. This epochal shift would come about, he assumed, in part because of his own métier. In this 'artificial world', 'climate and weather are actively controlled', genetically modified plants and animals 'are common in every landscape', and the genetic heritage of humanity itself enters 'a period of rapid divergence'. Evolution – even human evolution – would be determined by the market. The unconscious direction of natural history would cease and 'one would ask not how [ecosystems] evolved but why they were put there'.[146] Such a world dominated by the automatic subject would always be more unstable, dangerous, and irrational than neoliberal fellow travellers might realize. Instead, the geoengineer's dream seems to be the realization of Horkheimer and Adorno's horror: a 'wholly enlightened earth ... radiant with triumphant calamity'.[147]

Here, between Jennerite ecological scepticism and neoliberal Hegelianism, lies the crux of our intervention in the century-old socialist calculation debate. Neurath persuasively argued that socialism must be the conscious control of production and distribution, a political act that transforms the economy into the 'domain of the will'.[148] Mises and especially Hayek undermined Neurathian socialism through powerful epistemic critique, which diverted the Left into pseudorational market socialism. In response, we try to out-Hayek Hayek by arguing that nature is more unknowable than the market, and therefore far more deserving of our awe as an unconscious, decentralized, and unimaginably complex system. Neurath himself could not make this argument because he, too, embraced the humanization of nature as a socialist goal.[149] This blinded him to the fact that ecology has been neoliberalism's Achilles heel since the beginning.

Emphasizing our permanent ignorance of the biosphere allows one to reinterpret Hegel's philosophy of nature. While we see his aim of the complete humanization of nature as futile, that does not mean that there is no 'end of history'. The process of nature's humanization stops not when it is realized, but when our species comes to recognize that this process undermines the basis of human freedom. Climate change, emerging zoonotic diseases, and other environmental crises make a mockery of this pretence of control. To bring the humanization of nature to an end, the collective consciousness (*Geist*) must become aware of its own limits. As Neurath put it in a different context, 'rationalism sees its chief triumph in the clear recognition of the limits of actual insight.'[150] While the exact shape of the metabolic exchange between humanity and nature would certainly change over time, once this exchange is reduced to an ecologically stable size, then 'history' will more or less end.

Instead of the humanization of nature, much work in the future will be rewilding, which can be theorized as a kind of unbuilding of the world.[151] Compared to the Hegelian or Marxist labour of, say, turning a river into a canal or a meadow into a corn field, unbuilding would be the equally hard work of disentangling human consciousness from self-willed nature. This cracks open the concept of labour and our understanding of progress as we have known it since 1798, creating the possibility for a new admixture of work and leisure, as well as a new relationship between humanity and nature. The task of unbuilding makes clear that environmentalism isn't so much the idealization of 'pristine' nature (though it is vital to protect intact ecosystems) but the recognition that it is still possible to repair our broken world.

Half-Earth socialism will require labour in both its unbuilding and building forms – we'll need to install a lot of wind turbines – but there will be time for fun too. In this way, our conception of the end of history shares more with Hegel than one might expect. Alexandre Kojève's well-known interpretation can be given an ecological gloss:

> The disappearance of Man at the end of History, therefore, is not a cosmic catastrophe: the natural World remains what it has been from all eternity. And therefore it is not a biological catastrophe either: Man remains alive as animal in *harmony* with Nature or given Being. What disappears is Man properly so-called – that is, Action negating the Given, and Error, or in general, the Subject *opposed* to the Object. In point of fact, the end of human Time or History – that is, the definitive annihilation of Man properly so-called or of the free and historical Individual – means quite simply the cessation of Action in the full sense of the term. Practically, this means: the disappearance of wars and bloody revolutions. And also the disappearance of *Philosophy*; for since Man himself no longer changes essentially, there is no longer any reason to change the (true) principles which are at the basis of his understanding of the World and of himself. But all the rest can be preserved indefinitely; art, love, play, etc., etc.; in short, everything that makes Man *happy*.[152]

The aim for socialism is not Soviet-style Stakhanovite toil but *rien faire comme une bête*.

Although we have arrived at the end of history, that does not mean our project is complete. In the next chapter, we return to the cookshop of the future to put meat on the bones of Half-Earth socialism and see what the material embodiment of our philosophical principles might look like. We do this because we are inspired by Neurath's injunction for socialists to dream of new utopias to guide political action and practical planning. The Bavarian Soviet Republic, he thought, was defeated 'not at least because we lacked clear aims'. The problem was that 'Marxists killed playful utopianism ... paralysing the resolve to think up new forms.'[153]

That nature is ultimately unknowable does not mean we should not try to understand as much as we can about our beautiful, bewildering world. Our knowledge of nature will change, as will society and the technology it commands. This means that the conscious control of the economy and its unbuilding, too, will change over time. It is here that again Neurath can be our guide, for he was one of the first philosophers of science to recognize the protean nature of knowledge. He pithily expressed this insight by comparing scientists to 'sailors who on the open sea must reconstruct their ship but are never able to start afresh from the bottom.'[154] Creating socialism will not be easy, for below our half-built ship roils the merciless sea, but at least it is we who chart the course.

2

A New Republic

The food we eat masks so much cruelty. The fact that we can sit down and eat a piece of chicken without thinking about the horrendous conditions under which chickens are industrially bred in this country is a sign of the dangers of capitalism, of how capitalism has colonized our minds. We look no further than the commodity itself. We refuse to understand the relationships that underlie the commodities that we use on a daily basis.[1]

–Angela Davis

In what is perhaps the ur-utopia, Plato's *Republic*, Socrates incrementally added different trades to his perfect city, beginning with farmers, builders, weavers, and cobblers. In his initial plan, Republicans would eat an 'honest fare' of barley and wheat cakes, with the accompaniments of wine and music.[2] Glaucon, Plato's brother, protested against this bland diet, prompting Socrates to add 'salt, olives and cheese', as well as country stews of roots and vegetables with 'figs and chickpeas and beans' for dessert. Still unsated, Glaucon scorned this 'fodder' fit for a 'city of pigs'.[3] This time Socrates relented by adding meat to his imaginary menu, which required the integration of hunters, swineherds, and 'cattle in great numbers' into his Republic. The need for pasture forced the Republicans 'to cut out a cantle of our neighbour's land'. In turn, other city-states would attack the Republic if they, too, 'abandon themselves to the unlimited acquisition of wealth, disregarding the limit set by our necessary wants'.[4] Once the ideal of an irenic vegetarian society is abandoned, the nexus of land and meat emerged not only as a fundamental problem of political economy but also as 'the origin of war'.[5] Socrates, who rarely travelled beyond Athens' walls, believed 'the country places and the trees won't teach me anything'.[6] Yet he could not think through the problems of his Republic without considering humanity's relationship to nature because even a question as quotidian as diet unleashed a cascade of reciprocal effects on the economy and the ecosystem that undergirded it.

The term 'utopia' comes from a book of that title by Thomas More, who followed in Plato's philosophical footsteps some two millennia later. *Utopia* is about many things but, like *The Republic*, it confronts the tension between the desire and danger of dominating nature. Although Utopians 'count hunting the lowest, the vilest, and most abject part of butchery' and regard taking pleasure in death 'a cruel affection of mind', free citizens still ate meat and foisted the degrading act of slaughter onto the slave caste.[7] Instead of worrying that animal husbandry could spark foreign wars, More lambasted its role in the internal colonization of England. In his day, wool merchants turned 'meek and tame' sheep into monsters that 'consume, destroy, and devour whole fields, houses, and cities'.[8] With great acuity, More saw that England's nascent agrarian capitalism was a new political formation that allowed private property and money – a universal metric that 'beareth all the stroke' – to undermine the traditional ways of working the land.[9] Not content with the rents of yesteryear, entrepreneurial elites displaced peasants to expand sheep walks and profit from the booming wool trade. The burgeoning flocks of sheep were often afflicted by epidemics, which More interpreted as 'such vengeance God took of [the enclosers'] inordinate and insatiable covetousness, sending among the sheep that pestiferous murrain, which much more justly should have fallen on the sheepmasters' own heads'.[10] More may have acerbically attacked enclosures of previously common land, but like the meat-eating Utopians, he combined sensitivity with hypocrisy by being an encloser himself.[11]

More presciently discerned the ties between animal husbandry and early capitalism, a relationship that would become more obvious by the end of the century. In Shakespeare's *Richard II*, John of Gaunt laments how the 'scepter'd isle … is now leased out … like to a tenement or pelting farm', the means by which England made 'a shameful conquest of itself'.[12] Agrarian capitalism was not only rapidly remaking English society and its landscape, it was also linked to the 'origin of capitalist imperialism', as historian Ellen Meiksins Wood has argued.[13] In 1585, the Tudors expropriated land from Irish peasants and gave it to English settlers, who planted the first capitalist seed in foreign soil.[14] This process only accelerated after the ancien régime was defeated in the English Civil War. The victorious bourgeoisie removed any half-hearted Tudor-imposed restraints on enclosures and reinvigorated the colonization of Ireland.[15] Soon, much of the eastern half of the island was ruled by English and Scottish grazier-overlords, who profited handsomely from the export of live cattle to England.[16]

Colonization was justified by agrarian 'improvement', an argument made most famously by John Locke. Locke, whose theory of labour influenced Hegel's 'humanization of nature', believed that 'whatsoever [man] removes out of the state that Nature hath provided and left it in, he hath mixed his labour with it, and joined to it something that is his own, and thereby makes it his property.'[17] When the Lockean colonists of North America looked on the natives they sought to replace, they did not see a sophisticated constellation of complex and remarkably healthy societies – they saw people who did not work the land and therefore had no right to it. More, our ambivalent utopian antecedent, however, made this same insidious argument in *Utopia* nearly two centuries before Locke.[18] From Plato to More to our own time, it seems that nature and the animal question demarcates the limits of utopian possibility.

Whatever the shortcomings of the utopian tradition, its strength lies in the capacity to link food, land, ecology, and politics within a single analytical frame – an approach sorely lacking now. By contrast, none of the big three environmentalist 'solutions' today offer a similarly interconnected approach. Instead, mainstream environmentalists approach the environmental crisis as a set of discrete technical problems, addressable through piecemeal reform, while leaving the capitalist foundation of society untouched. These three demi-utopias are bioenergy carbon capture and sequestration (BECCS); greater nuclear power; and a colonial Half-Earth.

While neoliberals care little if their schemes of cap-and-trade or SRM actually work as long as the market's autonomy is preserved, mainstream environmentalists seem to really think that their demi-utopias might fix a broken planet. As we will see, these proposals have almost no chance of being implemented despite political concessions made to appease the rich and powerful. The frothy excitement over some green miracle cure such as 'fast-breeder' reactors or third-generation biofuels will soon fade alongside the hopes of reversing the crisis within capitalism.

By examining three of the leading proposals offered by the environmental establishment today, we show that they are insufficient not because of any technical shortcomings (though there are many) but because of their lack of utopian imagination. These schemes are simultaneously ambitious in scope and tepid in execution; they are divorced from any rigorous critique of contemporary political economy, as if the environmental crisis could be understood in isolation from the structure of the society that caused it. This strange separation of nature, economics, and politics was unavailable to the utopians Plato and More, whose epistemic holism we seek to revive. Try as they might, these mainstream environmentalists can't build a demesne wall between a technical solution to the future and the political-economic status quo of the present. Following our critique of BECCS, nuclear power, and Half-Earth, we outline our own utopia that imitates the bold scope of *The Republic* and *Utopia*.

BECCS: I'm a Loser, Baby ...

If one follows climate politics at all, BECCS seems to be everywhere except in any physical form. The concept behind the technology is simple enough: burn biomass to run a power plant, capture the CO_2 emissions, and sequester them underground. The biomass could come from the scraps of forestry and agriculture, plantations of specially developed trees, quick-growing grasses, and perhaps even algae one day. In the 2000s, BECCS' simpler technological forebear 'carbon capture and storage' (CCS) flopped because carbon prices were far too low to justify the installation costs.[19] Despite this troubling precedent, BECCS has recently become central to climate politics since the Paris Treaty of 2015. Small vulnerable island nations succeeded in including the target of 1.5°C warming in the treaty, which sparked an academic cottage industry in energy pathway simulations.[20] It was soon apparent that, barring rapid social change, the 1.5°C goal required not only carbon-neutral technologies like CCS but also carbon-*negative* ones like BECCS (i.e., the latter burns plants which have already absorbed carbon from the atmosphere). BECCS is the carbon equivalent of having your cake and eating it too.

The need for carbon-neutral technology becomes apparent when one considers humanity's 'carbon budget': the amount of carbon that can be released before climate change becomes catastrophic.[21] The budget is absolute rather than relative, because the atmosphere works somewhat like a bathtub – if it's already full, even a trickle of water will still cause it to spill over. Because so little of the carbon budget remains, restraining warming to 1.5 or even 2°C requires a precipitous fall in emissions. Carbon pollution will have to contract 7.6 per cent every year in the 2020s to achieve the 1.5°C target, surpassing even the 5.5 per cent fall in emissions during the pandemic-induced recession of 2020.[22] Soon after the IPCC published *Global Warming of 1.5 °C* in 2018, Carbon Brief reported that 'virtually all' the models used in the special report relied on negative emissions because of deficits in the carbon budget.[23] BECCS is appealing to modellers because it is one of the few ways to get their simulations to work. The only problem with the seemingly simple solution of BECCS is that there are no facilities in operation anywhere, nor does it seem likely that they will be built en masse anytime soon.

While the Paris Treaty may have spurred interest in BECCS, the concept dates back to the early 2000s. Kenneth Möllersten and David Keith – the Leibnitz and Newton of neoliberal science – both claim the hollow crown of BECCS' authorship. Twenty-two years ago, Möllersten was a doctoral student in chemical engineering studying how Swedish forestry could use BECCS to earn carbon credits from cap-and-trade programmes.[24] He travelled to the University of Cambridge in 2001 to present his research at the 12th Global Warming International Conference and Expo, where he met Michael Obersteiner of the International Institute of Applied Systems Analysis (IIASA), one of the original hubs for global environmental modelling (of which we will learn more in the next chapter). With Obersteiner and others, Möllersten wrote an influential report and article on BECCS and climate risk. Despite the high cost of BECCS, they argued, it was a flexible technology that could be deployed 'in situations of increased hazard'.[25]

Keith (the geoengineer we encountered in chapter one) wrote a sceptical editorial on BECCS in 2001. In 'Sinks, Energy Crops and Land Use', he argued that the technology aggravated land's 'essential scarcity' by competing with wild areas and agriculture.[26] A BECCS-based policy, he warned, 'must confront a still harder and more value-laden question, how much land ought we to spare for nature?'[27] Surprisingly, he seemed to think that rewilding would be a better way to sequester carbon (he remains sceptical of BECCS, but now opposes rewilding too).[28] Keith's scepticism has proven prescient. A study in 2018 found that if a BECCS programme were constrained by the nine essential 'planetary boundaries' (such as fresh water, biodiversity, etc.), then it would sequester a measly 0.5 per cent of annual emissions.[29] An effective BECCS programme that could sequester several gigatonnes of carbon would need at least 350 million hectares – an area larger than India.[30] At this scale of deployment, BECCS would actually *increase* global deforestation and exacerbate the Sixth Extinction.[31] Ironically, destroying so much habitat might make BECCS a net *source* for greenhouse gases.[32] So much for negative emissions.

Such a poor prognosis leads one to wonder why BECCS has become dominant in present-day climate policy. Again, the perspicacious Keith predicted that only after the failure of 'CCS, an expansion of nuclear, or a drastic cut in the cost of non-fossil renewables' took place would circumstances 'dictate the large-scale use of biomass energy'.[33] This is more or less what has happened. Keith might have guessed that there was always going to be little investment in BECCS because of its exorbitant cost; to keep warming limited to 2°C, BECCS would need to be a green tax of 3 per cent of global GDP from 2030 onwards.[34] In light of this, Keith's decision around this time to focus on SRM rather than BECCS made sense, because the farmer's cheapness always made it a more likely option. This is not to say that BECCS has nothing to do with neoliberalism; after all, Möllersten was inspired by cap-and-trade in the first place. The neoliberals are never wedded to any single solution but always have a panoply of policies to fill up the public sphere.[35]

While Half-Earth is supported by the relatively small community of conservationists, and the expansion of nuclear power is championed by high-profile environmentalists, BECCS seems to have only climate modellers as advocates. Their interest in the technology seems to be more technical than political, in that BECCS is a figment produced by their models rather than a real solution to climate change. Without a vigorous and radical climate movement, there is nothing keeping scientists politically grounded. Trying to solve the problem of climate change within the conservative parameters of climate modelling is a pointless exercise, which is perhaps why some members of the IPCC have become surprisingly radical, as we saw in the last chapter.[36] Utopias are meant to free one's conception of the possible, but demi-utopias like BECCS are fictional futures imagined to safeguard the status quo.

Fukushima, *Mon Amour*

Nuclear power has followed BECCS to the top of the climate debate less because of its ubiquity in climate models than because of the power and prestige of its advocates in the environmental movement, including James Hansen, Michael Shellenberger, George Monbiot, Stewart Brand, and James Lovelock. Their case for embracing nuclear power, which hitherto has been the bugbear of rank-and-file greens, depends on three assumptions: nuclear power is safe; nuclear plants provide 'carbon-free' power; and there are promising new nuclear technologies such as 'fast-breeder' reactors. None of these claims holds up under scrutiny.

One of the more sympathetic pro-nuclear greens is the former NASA scientist James Hansen. He is best known for being the first to present climate change to a broader public during his congressional testimony in 1988; more recently, it seems, he gets arrested all the time protesting at the White House.[37] Yet because of his pro-nuclear position, Hansen co-authored op-eds with Shellenberger (a conservative think-tanker) and Ken Caldeira (a geoengineer and collaborator of Keith's).[38] To decarbonize the energy system, Hansen advocates a vast and rapid roll-out of new reactors that would dwarf the current total of 440 worldwide. His scheme entails building 115 new reactors *every year for thirty-five years*.[39] Such a Herculean undertaking would reap surprisingly meagre results, as it would provide only enough power to supply the global electrical sector (roughly a fifth of total energy consumption).[40]

Rather than countenance changing how the Northern bourgeoisie (the social base of the environmental movement) lives, leading environmentalists put their trust in a technology as dangerous as nuclear power. It has been estimated that there is currently a 50 per cent chance of a disaster on the scale of Fukushima or worse every sixty-two years, and of Three Mile Island every fifteen.[41] If Hansen got his way and thousands of new reactors were built, then there would almost certainly be two, three, many Fukushimas (to paraphrase Che Guevara). A critical part of the pro-nuclear greens' argument is that there have been relatively few fatalities from nuclear accidents, even large ones like Fukushima and Chernobyl. According to Hansen and his peers, these nuclear meltdowns killed perhaps a few dozen people in Ukraine in 1986, while nobody died in Japan in 2011.[42]

This is nonsense. In 2006, the World Health Organization finally abandoned this ridiculous charade and raised Chernobyl's estimated death toll from 54 to 9,000.[43] Since then, the estimates have risen precipitously. That same year, a study carried out for the European Parliament offered the range of 30,000 to 60,000 deaths, while in 2019, historian Kate Brown argued there were between 35,000 and 150,000 deaths *at a minimum*.[44] Compared with Chernobyl, where the power plant was engulfed in flames for forty days, the disaster at Fukushima seemed less out of control, and only three deaths have been officially recorded in the years since the disaster. Again, this is almost certainly an underestimate. Two to five times more caesium-137 was released at Fukushima than at Chernobyl.[45] Caesium-137, like the better-known isotope strontium-90, easily lodges itself in the human body, where it can cause radiation poisoning and cancer. The estimate of 1,000 excess cancer deaths seems more realistic.[46]

Given the size, secretiveness, and strategic importance of nuclear industries, it is hard to hold them to account even when they fail. Cleaning up Fukushima is predicted to cost up to \$736 billion and last forty years.[47] It took eight years to construct a bespoke robot able to survive the conditions of the disaster's epicentre, and even then, it has merely made contact with the 'corium' – the magma-like amalgam of concrete, uranium, and the reactor itself.[48] The company that ran the Fukushima plant, TEPCO, was caught lying when it said during the early days of the crisis that the problem was only minor 'core damage' rather than a full meltdown. During the trial held on this self-confessed 'cover-up', the judge leniently agreed with the defendant that 'it would be impossible to operate a nuclear plant if operators are obliged to predict every possibility about a tsunami and take necessary measures'. This ignores how the company's own in-house models showed three years before the disaster that they were underestimating the risk of a tsunami.[49] In the end no one was convicted, but as an act of contrition TEPCO's president imposed upon himself a 10 per cent pay cut for a month.[50]

There is good reason to be sceptical of the pro-nuclear environmentalists' second claim of nuclear power being 'carbon-neutral'. There is a wide range in estimates of nuclear power's carbon impact because few agree on how much carbon is released during the mining and processing of uranium, decommissioning of reactors, and permanent storage of toxic waste.[51] This is why these assessments vary from 1.4 to 288 grams of CO_2 per kilowatt-hour (gCO_2/kWh), with a mean of 66 gCO_2/kWh.[52] By comparison, solar power and wind can have a carbon footprint as low as 1 gCO_2/kWh (depending on technology, longevity, and location) with the mean of 49.9 gCO_2/kWh for solar and 34.1 gCO_2/kWh for wind.[53] Most climate models for 2°C of warming assume that a reduction to 15 gCO_2/kWh is necessary for all power generation by 2050, which seems feasible for wind and solar but difficult for nuclear.[54] Nuclear's problem is that the mining phase of its life cycle is set to become ever more carbon-intensive. Most power plants today rely on uranium ore of 0.15 per cent purity, which translates to 34 gCO_2/kWh. Yet once available ore quality declines to 0.15–0.01 per cent, carbon emissions jump up to 60 gCO_2/kWh.[55] While this is not expected to happen until the 2060s at current rates of extraction, this cliff will arrive much sooner if a build-out occurs on the scale Hansen advocates.[56] If there is good reason to think that nuclear power lacks the low-carbon profile that justifies its inclusion in the global energy mix, then the heart is ripped out of the pro-nuclear position.

Uranium's impending scarcity is one reason why the pro-nuclear greens extol the virtues of 'fast-breeder' reactors, their third argument for a nuclear renaissance.[57] The moniker 'fast-breeder' refers to a reactor that uses nuclear waste as fuel. Conventional reactors use relatively scarce uranium-235, which leaves 'depleted uranium' (U-238) as a by-product, while fast-breeders use U-238 as fuel and produce plutonium-239 as waste – a fissile material that could be used for fast-breeders or H-bombs. Hansen and company portray fast-breeders as a high-tech, low-carbon, limitless energy source that produces little toxic waste, but overlook the fact that these reactors simply don't work. This is not for lack of trying, as governments around the world have spent $100 billion over seven decades to commercialize the technology, to no avail.[58] One of the more recent developments is the strange bipartisan push in the United States, where then president Donald Trump and vegan senator Cory Booker joined forces to build a new breeder in Idaho designed by GE Hitachi and Bill Gates' company TerraPower (Gates is a vocal supporter of breeders and SRM).[59] Right now, only two breeder prototypes remain in operation, and both are in Russia. Yet they don't even run on plutonium – defeating the whole point of their 'closed' fuel cycle – and one of them boasts the record of operating continuously despite catching fire *fourteen times* in seventeen years.[60]

Conflagrations are common because breeders use liquid sodium as a coolant, an element that burns with air and explodes with water. As a result, breeders are frequently shut down for repairs, giving them a very low capacity factor (i.e., how often they run). Japan's $9 billion Monju prototype was typical in that it operated for only 250 days over its twenty-two-year existence because of sodium leaks, including a serious fire in 1995 that left the plant shuttered for fifteen years.[61] Only India and China are still building new breeders, but less to generate power than to harvest plutonium for H-bombs. India's reactor, however, has been stymied for decades by its unruly sodium coolant, with no end to the problem in sight.[62] It may be the case that the liquid sodium reacts with the carbon in the steel encasing it, leading to 'metal dusting' and leaks.[63] Although the proponents of breeders blame the failures on human error or teething troubles, there is a good chance that the technology is fundamentally flawed.

The nuclear turn in environmental discussions today is bewildering as a matter not only of technology but also of tactics. After all, environmentalists' pro-nuclear turn is a strange twist for a social movement that largely emerged through its critical engagement with nuclear science. In 1953, the RAND Corporation think-tank undertook Operation Sunshine (one of the first global environmental monitoring systems) to track the radioactive isotope strontium-90 spread by nuclear testing.[64] RAND's scientists set up forty-four monitoring stations in the US and forty-nine abroad, which allowed them to trace the links between 'human cells, plants, animals, landmasses, water systems, jet stream patterns, and the atmosphere with increasing precision'.[65] The best way to measure strontium-90, however, was in human bones – a dataset that could not be easily assembled legally. The project's surreptitious 'body snatching' came to light early in 1957, but the full extent of such crimes only became clear in 2001.[66]

Not willing to leave this research solely to the military-industrial complex, activists in the newly formed Greater St. Louis Citizens' Committee on Nuclear Information (CNI) began their 'baby tooth survey' in 1958. The CNI was a coalition of pacifists, Quakers, progressive politicians, and engagé scientists at the nearby Washington University. They published their first results in *Science* in 1961.[67] A year later, Rachel Carson wrote the seminal *Silent Spring*, which applied the lessons from strontium-90 to pesticides, each a substance that 'lodges in soil, enters into the grass or corn or wheat grown there, and in time takes up its abode in the bones of a human being, there to remain until his death'.[68]

The baby tooth survey was an effective lesson in coalition building. While the military researchers associated with RAND could steal skeletons, the civilians in St Louis had to make do with donated teeth.[69] To collect a sample size of 50,000 baby teeth, the CNI carried out teach-ins at churches, Boy Scout meetings, dental clinics, and libraries. Volunteer battalions disseminated a million tooth-survey forms and carefully catalogued the tens of thousands of mailed-in teeth.[70] This research revealed that strontium-90 levels in baby teeth spiked in 1954 and 1955, which closely correlated with the rise in nuclear testing. In early 1963, one of the CNI's scientists was invited to present these findings at a US senate hearing.[71] His testimony contributed to the 1963 Nuclear Test Ban Treaty, which lowered strontium-90 levels. Such victories are possible when politicized scientists and social movements unite.[72]

Rather than a pro-nuclear agenda, we need another baby tooth coalition both as a model for the environmental crisis today and to continue the fight against the still-dangerous nuclear establishment. After all, strontium-90 is not only released by nuclear tests but also by civilian plants, which bodes ill if Hansen ever succeeds in realizing his demi-utopia of thousands more reactors.[73]

A Wild Foundation

Ideas can be like cashews – while shelling, one must avoid the acid to reach the fruit. In the case of conservation, the acid is colonial Malthusianism and the fruit is the protection of thousands of species from extinction. Given that the main driver of the Sixth Extinction is land-use change, conservation is essential to our eco-socialist utopia. Our criticism of conservation's colonial past is hardly novel, but unlike other critics we delve specifically into the history of Half-Earth to try to salvage the idea for socialism.[74]

The history of Half-Earth is the history of the relatively small environmental groups that conceived and developed the concept. The heavy hitters of the movement, such as Greenpeace or the World Wildlife Fund, are nowhere to be found in this narrative. While E. O. Wilson popularized Half-Earth with his 2016 book of the same name, our narrative mainly focuses on prior environmentalist campaigns. Seven years before Wilson's book, the WILD Foundation appointed Harvey Locke, a Canadian conservationist, to head its 'Nature Needs Half' campaign because he had worked on the idea early in his career. Under his leadership, the Canadian Parks and Wilderness Society (CPAWS) decided in 2005 to fight for preserving half of Canada – an unprecedented target. To understand why Locke came to embrace this ambition, it is necessary to examine his association in the 1990s with a fringe environmentalist group called the Wildlands Network.

This organization had been set up in 1991 by Earth First! co-founder Dave Foreman, the biologist Michael E. Soulé (a student of Paul Ehrlich), and Douglas Tompkins (the fashion mogul behind The North Face and Esprit).[75] Rather than engage in direct-action campaigns like Earth First!, the Wildlands Network focused on drafting blueprints of a rewilded North America. In 1992, one of the group's leading members, Reed Noss, made the case in *Wild Earth* (their house magazine) for rewilding 50 per cent of the continent. His reason for such an ambitious target was the need to 'restore viable populations of large carnivores and natural disturbance regimes'.[76] Notably, this article did not draw on Wilson's work on biogeography, but Reed and Soulé eventually did so six years later.[77] At the time, Wilson modestly called for 'expand[ing] reserves from 4.3 per cent to 10 per cent of the land surface'.[78] Locke, who worked closely with the Wildlands Network and even joined its board of directors, organized the Canadian side of the Wildlands Network's 'Yellowstone to Yukon' campaign. This was to be the first of several 'wildways' connecting parks across the continent that would allow migratory species and apex predators to roam freely. In the early 2000s, the three streams of the inchoate Half-Earth movement converged when both Locke and Wilson contributed essays for *Wild Earth*.[79]

The Wildlands Network's approach to conservation was bold and prescient, but that does not excuse the group's toxic politics. Foreman succeeded in putting an anti-immigrant resolution to a membership-wide vote at the Sierra Club in 1998. (It was defeated.) In the 2000s, he received funding from the far-right Weeden Foundation to co-ordinate the Wildlands Network's activity with the anti-immigrant group Apply the Brakes (where he also held a leading position).[80] In 2015, Foreman published a second edition of his screed *Man Swarm: How Overpopulation Is Killing the World*, which cited 'research' produced by John Tanton's white-supremacist think-tank Reich.[81] In the foxhole of the environmental crisis, one is either a Marxist or a Malthusian.

While Locke is certainly more tolerant than Foreman, he plays a leading role in the WILD Foundation – another group with a dark and frankly bizarre past.[82] Initially known as the International Wilderness Leadership Foundation, the WILD Foundation was founded in 1974 by three men who each represented an archetype of the conservation movement. Ian Player was the naturalist, who worked as a game ranger in South Africa and pioneered the capture and transport of large mammals as a conservationist tactic. Robert Cleaves was the soldier, whose private air force, WILDCON, spurred the militarization of conservation. An admirer of Rhodesia, Cleaves represented the US government in Harare when Ian Smith's white supremacist government surrendered to Robert Mugabe's Zimbabwe African National Union in 1980. WILD's third founder, Laurens van der Post, was a mystical con man. Pretending to be Carl Jung's disciple and an expert on the San people, Post managed to charm his way into the circles of Margaret Thatcher and Prince Charles (who even asked him to become Prince William's godfather). After Post's death in 1996, one of his early admirers went through his papers to write a biography but discovered that everything from his war record to his Dutch aristocratic lineage was fraudulent.[83] Even before Post's secrets tumbled out, wags had christened the fishy Jungian 'van der Posture'.[84]

The roles of the mystic, naturalist, and soldier are more complementary than they might appear. Post inculcated a Jungian spirituality in Player and other leaders in the WILD Foundation (including Locke and the current president, Vance Martin). Player went so far as to found the Cape Town Centre of Applied Jungian Studies. Player depended on Cleaves' air support to patrol his parks, just as he needed his right-hand man Nick Steele (one of the 'hard men' of South African conservation) and his troop of former Rhodesian mercenaries on the ground. While Player cut a more sympathetic figure than Cleaves, he shared much of Cleaves' politics. Steele recorded in his diary how Player confessed that if Cuba defeated South Africa in the war in Angola, '[the Africans] would seek their retribution on us [and] I'd rather commit suicide than face that'.[85] These three facets of the environmental movement provide belief, legitimacy, and force.

Although the WILD Foundation now plays down its links to the South African government, the organization provided vital support to the embattled apartheid regime. The WILD Foundation's international conferences lessened the country's isolation by inviting well-known environmentalists like Stewart Udall, Gro Harlem Brundtland, and Maurice Strong.[86] Furthermore, Player's new parks in KwaZulu-Natal served as a buffer against African National Congress (ANC) forces in Mozambique.[87] To get support for its parks, the WILD Foundation allied with the quisling leader Mangosuthu Buthelezi and his Inkatha Freedom Party (IFP).[88] This conservative Zulu nationalist organization was supported with money and arms by Pretoria to further intra-African strife that left 15,000 dead.[89] One of the WILD Foundation's benefactors, gambling magnate John Aspinall, gave the IFP 4 million rand during the 1994 election, and urged an IFP crowd to 'sharpen their spears and fall on the Xhosas [the dominant ethnic group in the ANC]'.[90] Clearly, the WILD Foundation preferred the ersatz traditionalism of the IFP to the ANC's modern, internationalist socialism.

Compared to bloodthirsty conservationists like Foreman and Cleaves, Wilson appears mostly harmless. He is a bogeyman for the Left because of his book *Sociobiology*, which naturalized sexual and cultural differences.[91] Apart from this admittedly reactionary research programme, Wilson is a centre-Left Democrat who thinks that policy nudges and the generosity of enlightened philanthropists suffice to achieve planetary conservation.[92] Regardless of his politics, his scientific work on island biogeography has withstood more than five decades of scrutiny. There is simply no way to stop the Sixth Extinction other than expanding nature preserves (under Indigenous leadership wherever possible). Still, socialists are right when they criticize conservation for burdening poor and Indigenous people.[93] The solution is that Half-Earth must be socialist, not that socialism doesn't need Half-Earth.

Like BECCS and nuclear power, the colonial Half-Earth is a demi-utopia that luckily doesn't have a hope of being implemented. Profound transformations of the carbon cycle, the energy system, and biodiversity will have a better chance at realization when combined with planetary in natura economics and a broad liberatory coalition. This means that conservationists need to break with the guardians of the capitalist status quo – the plutocrats, charlatans, and mercenaries – and join its gravediggers instead.

Forces of Nature

Our aim in this section is to understand why capitalism produces more and graver ecological problems than any other social form in human history, so that socialism can better avoid them. To this end, it is helpful to return to the place and time of capitalism's birth, five centuries ago in the English countryside. It was then that capitalism engendered not only the self-conscious genre of utopianism but also a new relationship to nature.

To understand this relationship, we will use Marx's concept of the 'natural forces' (*Naturkräfte*) as a physical counterpoint to the more abstract idea of the humanization of nature that we analysed in the first chapter. These forces include not only the 'impulse' setting in motion the 'water-wheel from the descent of water down an incline, [and] the wind-mill from the wind', but also 'human exertion'.[94] While labour power is the only natural force that produces surplus value for capitalists, capitalists are otherwise indifferent to which natural force they employ in production.[95]

This perspective helps explain why capitalism emerged from animal husbandry as well as its relationship to the Industrial Revolution. Replacing peasants with flocks of sheep and a few shepherds was a sort of proto-mechanization that increased relative profit by increasing labour productivity. 'The raising of sheep required fewer labour inputs per acre than the growing of grain,' economist William Lazonick explains, '[which] directed the use of land away from production of the fundamental means of subsistence to production for the market and for profit ... [and] due to its land-intensive technical requirements, it separated many producers from the means of production.'[96] Marx saw the eighteenth-century breeder Robert Bakewell as representative of capitalist agriculture. Bakewell's family belonged to the new caste of 'improving' farmers, whose ancestral home was among the first to be enclosed (as far back as More's lifetime).[97] Before Bakewell, domesticated sheep took five years to mature, but he managed to cut this down to a year by reducing 'the bone structure ... to the minimum necessary for their existence', Marx observed.[98] Marx recognized that in a capitalist society there was no difference between a sheep breeder like Bakewell and a locomotive manufacturer who adopted new machine tools, because both were capitalists who sought to increase profits by reducing turnover.[99] Blind capital sees little difference between animal and machine; both are instruments to raise labour productivity.

The sixteenth-century boom in wool was eventually surpassed by cotton, an industry that made use of water-powered mills. The first recognizably modern textile mill was built in 1771 in Cromford by Richard Arkwright. Whereas Arkwright's previous enterprise in Nottingham was powered by horse muscle (3–4 horsepower), the Cromford factory was the first to rely on water power (10 hp), while waterwheels a generation later were much more powerful (100 hp).[100] The shifts from one kind of natural force to another – labour power, animal muscle, water, coal – did not follow a secular trend but were chosen by capitalists based on their advantages at any given historical conjuncture. Water was cheap and powerful, besting coal on both counts for almost three generations after Arkwright's mill was built. The shift from water to coal occurred because capitalists sought to reassert control over unruly workers. It was difficult to attract labour to mills in the isolated river valleys, and rampaging Luddites could easily destroy a factory there before the army could be called in. Coal offered the possibility of placing factories in towns, near barracks and a pliable industrial reserve army of proletarians, many of whom had been driven from the countryside by enclosures.[101] In the twentieth century, petroleum undergirded an even more flexible energy system that workers struggled to control.[102]

The future, however, may herald a return of animal power and what sociologist Kenneth Fish calls the 'agriculturalization of industry'.[103] Fish sees genetically modified organisms (GMOs) such as 'spider-goats' – which produce arachnid silk in their udders for things like bulletproof vests – as the purest encapsulation of capital's relationship to nature, that is, the redirection of natural forces to further capital's self-expansion. According to Fish, labour under capitalism has changed little since shepherds replaced farmers in More's time: most work takes the form of 'eco-regulation', in which natural forces are guided by human attendants.[104] This is why Marx describes the factory as an 'entirely objective productive organism' (*ganz objektiven Produktionsorganismus*) where the worker becomes a mere appendage to the machine.[105] Notably, Fish emphasizes that eco-regulation applies equally to labour carried out on the farm and in the factory, which is why 'for all the technological mastery marked by the coming of the machine, then, the significance of the factory for Marx lies in how it approximates a living organism, that most natural of beings.'[106]

Over the last century, Bakewell's techniques have been taken to an extreme in pursuit of greater labour productivity.[107] The growth rate of 'broiler' chickens increased by 400 per cent between 1957 and 2005.[108] Between 1950 and 2020, annual milk production per cow grew from 2,400 litres to 10,600 litres.[109] Sustaining so many animal-machines requires immense resources, which in turn threatens wild species with extinction. Animal husbandry takes up 4 billion hectares – 40 per cent of Earth's inhabitable land.[110] It is no wonder, as a recent study has found, that 'animal product consumption by humans (human carnivory) is likely the leading cause of modern species extinctions'.[111] Capitalist agriculture has created a world where 60 per cent of the total terrestrial mammalian biomass is livestock; only 4 per cent is wild mammals, while humanity composes the remaining 36 per cent.[112] Because of the sheer bulk of domesticated animals, some experts argue for including the *respiration* of such artificially abundant life as carbon pollution.[113] As Fish might argue, these animals should be seen as living factories no different from smoke-belching industrial factories. Capitalism, born in the countryside, must die there if there is to be any hope for a new, ecologically stable socialism to take its place.

While we have stressed that capital is indifferent to the various forms of *Naturkraft*, there is an important distinction between 'flows' and 'stocks'. Renewable systems rely on flows of energy, whose source generally is solar radiation (tidal and geothermal systems produce modest power). It is a cheap and plentiful source of energy, but solar flows are by definition variable and dispersed and thus have a low 'power density', which is measured in watts per square metre of land (W/m^2).[114] Solar and wind power produce about 5 to 10 W/m^2, while biofuels produce a miserly 0.5 W/m^2. By contrast, fossil fuels represent concentrated stocks of energy and thus have extremely high power densities. The richest petroleum deposits in Saudi Arabia harbour 40,000 W/m^2, and even shabby ones like Canada's tar sands still boast 1,100 W/m².[115] The concept of power density provides some unity and insight into our preceding discussion. Crops generally have a low power density, which is why the livestock industry and BECCS devour land. Indeed, the low power density of BECCS and renewables is one reason why some greens support nuclear power. While uranium is a stock rather than a flow resource, the pro-nuclear greens overlook how protective glacis and cooling lakes can significantly lower power density: the doomed Fukushima Daiichi plant produced 1,300 W/m², but the Wolf Creek facility in Kansas has a modest power density of 30 W/m².[116]

If Half-Earth socialism renounces the use of stock energy sources like fossil fuels and nuclear power, then land scarcity will become a major economic and ecological constraint. Just as in Plato's day, the territorial demands of the livestock industry limit the utopian imagination, but unlike the ancient Greeks, we must also convert a power-dense energy system into a power-sparse one, and at the same time prevent an extinction event of an extent unimaginable in a pre-capitalist era. If the essence of the problem is the humanization of nature, what does the solution of unbuilding look like in practice?

Half-Earth in Havana

The goals of Half-Earth socialism are simple enough: prevent the Sixth Extinction, practise 'natural geoengineering' to draw down carbon through rewilded ecosystems rather than SRM, and create a fully renewable energy system. Realizing each of these aims requires large expanses of land, which is why we will see again and again that utopia is threatened by the Earth-eating livestock industry. Luckily, these three goals are complementary. Greater biodiversity increases the carbon sequestration potential of an ecosystem, while a decarbonized and vegan agricultural system will free up space for rewilding and renewable systems. An eco-socialist future is salvage-able – even at this late stage of the environmental crisis – but it requires Neurathian planning so one can discern the other-wise opaque workings of the economy and envision a utopian alternative.

Rewilding means not only allowing natural forests and grasslands of native species to replace pasture but also returning wild animals to these ecosystems. Healthier, more biodiverse ecosystems sequester more carbon than simplified ones – including the gigantic BECCS plantations imagined by some modellers.[117] Much of the livestock population today is genetically similar ruminants (e.g., domesticated cows, sheep, and goats), whereas the lifeblood of a healthy ecosystem is large nonruminant herbivores, such as the white rhinoceros, wildebeest, Bactrian camel, Przewalski's horse, African wild ass, and kulan.[118] Due to their different digestive tracts, these animals also produce much less methane than domesticated ruminants.[119] The restoration of large frugivores (fruit-eating animals), such as tapirs and Asian forest elephants, could increase the carbon sequestration capacity of tropical forests by 10 per cent.[120] Predators matter too. If Canada's wolves returned to their former glory, their predation of moose would create a healthier boreal forest that could potentially cancel out all of Canada's current carbon emissions.[121]

Oceans are vital to Half-Earth too. The oceans shelter half of all life and sequester about 30 per cent of global carbon emissions – some two gigatonnes a year. Although they occupy only 0.2 per cent of the seafloor, seagrass ecosystems absorb as much as a tenth of all the organic carbon absorbed by the ocean every year.[122] They are also urgently in need of pro-tection, as they are one of the most endangered ecosystems, facing an annual rate of depletion of 7 per cent.[123] Whales deliver plankton from the ocean's surface to its depths through everyday acts of eating, diving, and excreting.[124] The bodies of living whales contain as much carbon as the forests of Rocky

Mountain National Park. In death they entomb an estimated 30,000 tonnes of carbon every year (and up to 160,000 if their populations were allowed to recover) by sinking to the ocean floor. Some marine biologists have thus concluded that 'the impact of rebuilding stocks of fish and whales would be comparable to existing carbon sequestration projects'.[125] Yet marine populations are down by 49 per cent since 1970 (itself hardly a halcyon era for sea-life), while fishing robs whales of food or entangles them in trap-lines.[126]

Wilson's Half-Earth should be seen as a kind of natural geo-engineering. His plan identified thirty biomes ranging from the Brazilian *cerrado* to the Polish-Belarusian Białowieża Forest that would be the heart of Half-Earth.[127] These would eventually be stitched together (much like the Wildlife Network's 'wildways') to create an interconnected mosaic spanning half the globe. Rainforests can sequester 200 to 650 tonnes of carbon per hectare (tC/Ha), while a Californian redwood forest can contain 3,500 tC/Ha.[128] These are among the highest rates found on land, and the preservation and expansion of such forests should be the centrepiece of any climate policy. Climate scientist Ulrich Kreidenweis estimates that reforesting 2.6 billion hectares (i.e., two and a half Canadas) could entomb 860 gigatonnes of CO_2 by 2100.[129] It is unlikely that such a bounty of 'negative emissions' could be matched by other sequestration technologies like BECCS.[130]

The easiest – and perhaps only – way to achieve large-scale reforestation and feed the world at the same time is through widespread veganism. In a 2016 study, modellers ran 500 scenarios based on diet and global reforestation and found that while all of the vegan pathways and most of the vegetarian ones (94 per cent) were possible, only 15 per cent of the rich-world diet scenarios succeeded.[131] Kreidenweis also found that if mass afforestation were carried out without reducing meat consumption, then food prices would jump globally by 80 per cent by 2050 and by 400 per cent by 2100.[132] Whether organic (and thus low-carbon) agriculture can feed a growing global population has long been debated, and the verdict appears to be yes, but such a system cannot produce much meat or dairy.[133] This isn't surprising; the livestock industry requires vast monocrops of soy and maize that organic agriculture cannot easily replace. David Pimentel and his co-authors found that yields are nearly equivalent for organic vegan agriculture and industrial agriculture on a year-by-year basis.[134] Of the major crops, it is only maize that is the victim of the 'rotation effect', because it can only be planted every year on the same field through determined use of fertilizers and pesticides.[135] Without the capitalist pressure to decrease turnover à la Bakewell, growing other crops on the same field is hardly a problem. Smaller fields (with more hedges and the like) and fewer pesticides also allow organic farms to host significantly more biodiversity than conventional ones.[136]

Half-Earth socialism's third aim, of constructing a completely renewable energy system, only makes the problem of land scarcity more pressing. The high power density of fossil fuels, as well as renewables' small share of the total energy mix, means that currently only 0.5 per cent of US territory is occupied by its entire energy system.[137] What would a completely renewable energy system look like in terms of land use? Energy expert Vaclav Smil estimates that such a system would take up 25 to 50 per cent of the US land-mass, while rich and densely populated countries like the UK would have a ratio approaching 100 per cent.[138] Although it is the most frequently discussed facet of energy policy, converting the electricity sector to renewables would be the easy part, but that represents only about a fifth of total energy production. Smil estimates that the 320 GW of US fossil-fuel electrical production could be replaced by solar and wind power infrastructure that would take up only 22,000 km^2 (an area about the size of New Hampshire).[139] This will be much easier than the larger and trickier sectors of industry (700 GW) and transport (1,100

GW). Furthermore, some commodities lack good fossil-fuel substitutes, such as jet fuel, coke, and cement clinker. Biofuels can perform some of these tasks, but it's best not to rely too much on these land locusts. Indeed, biofuels are the main reason why Smil's land-use estimates are so high. Hydrogen might become a better alternative to biofuels, but we can't depend on its adoption anytime soon.[140] (We will talk more about the challenges posed by different energy sources in the next chapter.)

One way to save land would be to produce less energy, which is why Half-Earth socialism embraces quotas. The exact number can be debated, but we admire the target of the 2,000-Watt Society. This brainchild of Switzerland's Federal Institute of Technology proposes a global energy consumption converging at 2,000 watts per person, which would require severe cuts in the rich world, while allowing growth in poor countries.[141] This in itself would go a long way to level inequalities in global living standards. Today, an average US citizen uses 12,000 watts, a western European 6,000 watts, and an Indian just 1,000 watts. Indeed, much of humanity would be better off in absolute terms under Half-Earth socialism than under the current capitalist system. By instituting quotas, the energy sector would take up less land, so that even small, densely populated nations like the UK would have enough space for renewable energy, Half-Earth, and vegan agriculture. Yet it's impossible to imagine such a programme being carried out within capitalism, for the automatic subject would soon enough champ at the bit of such restrictions. This is why conscious control of the economy is necessary.

An economic system resembling Half-Earth socialism can actually be found in recent history: Cuba's Período Especial. In 1990 the Soviet Union stopped subsidizing petroleum imports to its socialist allies, and with little hard currency to buy it on the world market, Cuba had to decarbonize almost overnight. At the time, Cuba's model of industrial cash-crop production left it more reliant on fossil-fuel inputs than US agriculture.[142] Getting by without petroleum or petroleum-based products (e.g., fertilizers and pesticides) forced the largest and most compressed experiment in organic and urban gardening in history. Soon, there were 26,000 urban gardens in Havana alone, allowing the city to satisfy its own requirements for fresh vegetables.[143] The government bought more than a million bicycles from China to replace the idling buses and cars. Eating less meat and more vegetables, combined with pedalling or walking to work, led to improved health in the general population.[144] Despite an economic contraction and the tightening of the US embargo, universal health care and education were maintained and many indices even improved.[145] Cubans cultivated less land more intensively, returning about a third of farmland to wilderness.[146] This has helped Cuba maintain its incredible biodiversity (it is listed among Wilson's top thirty biomes) and led the World Wild Fund for Nature to recognize it as the world's *only* 'sustainable' country.[147] Cuba suffers less from common environmental problems such as invasive species, 'colony collapse disorder', and plastic pollution.[148] Cuba's transition to an ecological society has been difficult, to say the least, but if this poor, isolated island could refashion itself during a severe economic crisis into a novel form of eco-socialism, then no rich country has an excuse for inaction.

Rewilding, energy quotas, and widespread veganism are effective, simple solutions that are available right now, even if they might struggle to find public support initially. Indeed, the example of Cuba's Período Especial may repel as many as it attracts. They become, however, more appealing when compared to the insufficient solutions of the Half-Earth of Wilson and the WILD Foundation, nuclear power, and BECCS, which fail to trace all the interconnections between ecological and economic goals. More and Plato were well aware that utopian thought is akin to double-entry bookkeeping, because changing one side of the ledger, such as the addition of a carnivorous diet, requires a change on the side of the political economy. To the extent that they try to make these sums work, environmentalists have been willing to cook the books. We offer an honest reckoning of Half-Earth socialism because we believe that a feasible utopia is one where its costs are democratically appraised rather than hidden by the pseudorational measure of money.

Politics Is Like Pulling Teeth

The Half-Earth coalition must be a broad one. There should be animal-rights activists and organic farmers there, as well as socialists, feminists, and scientists. The existing Half-Earth project does not have a large social movement supporting it, which is why Wilson and the WILD Foundation have relied on philanthropists and reactionaries. BECCS exists only in the clouds of supercomputers, and its technocratic admirers might fill a few lecture halls, but not the streets. Yet the most powerful rallying cry available to the environmental movement has historically been the fight against nuclear power. By embracing such technology, the pro-nuclear greens are essentially arguing against the environmental movement's most popular political position, in return for little gain, as we have seen. In many ways, the pro-nuclear greens represent a betrayal of the environmental movement's history, and potentially its future too.

The nuclear question is one of the few issues that not only unites environmentalists with other social movements but is also capable of wide-scale mobilization. Compared to minority interests such as veganism or the Sixth Extinction, anti-nuclear activism commands a majority. It routinely wins referendums: Austria 1978, Italy 1987, Italy 2011, Lithuania 2012. The first action by Greenpeace took place in 1971, when a dozen activists sailed an old fishing boat to the remote Alaskan island of Amchitka to 'bear witness' to a nuclear bomb test there.

Indeed, the group's name shows how environmentalism and pacifism once went hand in hand. Germany's Green Party, one of the world's most successful, started out in 1979 as the electoral vehicle for the burgeoning anti-nuclear movement.[149] Fear of nuclear power has proven a powerful mobilizing force for green parties around the world, especially in the wake of disasters like Chernobyl and Fukushima. Looking at the former, a political scientist correlated radiation levels in Swedish towns and the share of the Green Party's vote in the 1988 election.[150] Just weeks after the Fukushima disaster, the German Green Party took power at state level for the first time, with 24 per cent of the vote in Baden-Württemberg.

The adopted pose of the pro-nuclear greens is the no-nonsense leader willing to make unpopular decisions for the greater good. Michael Shellenberger decries the 'ignorance' and 'ridiculous' concerns of nuclear sceptics ('nuclear waste is the best kind of waste').[151] James Hansen likens the belief that a fully renewable energy system is possible to continued faith in the 'tooth fairy'.[152] (Ironically, the tooth fairy was the anti-nuclear movement's strongest ally in the late 1950s.) In the same vein of ostensible realpolitik, the pro-nuclear greens cite renewables' low power density as a major reason for championing nuclear power. This argument only appears to make sense because they refuse to change other parameters in order to save land, such as abolishing the livestock industry. Indeed, Shellenberger argued on Fox News in 2019 that vegetarianism would have little effect on greenhouse gas pollution and that factory farming of livestock is the best way to protect the environment.[153] As we have seen in the case of Chernobyl's and Fukushima's death tolls and the fabled potential of breeder reactors, pro-nuclear greens' defence of omnivory is not the only instance where they dilute the truth to greenwash nuclear power.

Nuclear power, BECCS, and the Half-Earth as promoted by Wilson and Locke have more in common with SRM than with Half-Earth socialism. They are promoted as a partial cure without addressing either the environmental crisis as a whole or its capitalist drivers. Their very point is to preserve as much of the status quo as possible, but such political minimalism courts disaster. BECCS would accelerate the Sixth Extinction, and more nuclear power risks new Chernobyls, while Half-Earth will further conservation's role as a form of neo-colonialism. One could add that none of these demi-utopias has a chance of being realized, because their obsequiousness towards power fails to inspire the social movements needed to actually make the world better. Environmentalists think that their political moderation will be rewarded by the powerful, who will allow them to play a small role in shaping the future. Yet the elite and their neoliberal brain trust have no need to make concessions, when SRM is close at hand and the environmental movement is weakened by infighting. The powerful will only concede what is taken from them, and only a broad and radical movement can force such concessions.

When we consider the uninspiring demi-utopias on offer today, we have to wonder why it has become so difficult to conceptualize a society in all its facets as Plato and More could. A notable difference between More and Plato is that Plato imagined a society not so different from the one he lived in, yet More created a fictional island seemingly far removed from England. As the self-consciously utopian genre emerged concomitantly with capitalism, its fictional aspect is not an incidental detail but rather a manifestation of the separation in capitalism of the political sphere from the economic. A utopian socialist like More could only reconcile the two in fiction, in a way that was unnecessary for Plato because he lived in a more directly mediated society. More recently, even this utopian impulse has slackened, leaving thinkers and artists today able to foresee the apocalypse and little else. Mainstream environmentalists cannot even imagine a radically different society because they have no utopian tradition to draw upon, and thus are left with dreary technical solutions that inspire few. (Malthusianism, it seems, is the dark poetry that can enlivens their imagination.) We have tried to revive the utopian socialist tradition, but we hew closer to the more practically minded philosophers like Neurath and Plato. This is why, when we try our hand at utopian fiction as social engineering in the final chapter, we place it in Massachusetts, where we were living when we conceived this book.

Admittedly our utopia would still be constrained by scarcity, but true prosperity is measured in land, not dollars. We concede that some of Half-Earth socialism's attributes, such as energy quotas and giving up meat, might not appeal to all, but we think such sacrifices are more attractive than the three demi-utopias offered by the environmental movement. The environmental crisis is already well advanced, so it should not come as a surprise that there are only hard choices left to make. However, Half-Earth socialism is not merely a 'lesser evil'. One might see that there is much to gain by giving up control over a subsumed nature. While much value must be destroyed by relinquishing the land and sea that has been trawled, mined, and razed, there is a new wealth to be gained too. A bounty of beauty, safety, and stability will come from the thousands of species that will be protected, the gigatonnes of carbon sequestered, the promise of meaningful work and social security, for Half-Earth socialism will be a rich society too. Even if it means figs and beans for dessert.

3

Planning Half-Earth

Perhaps we are now at the beginning of a scientific study of utopias.

–Otto Neurath

For Leonid Kantorovich, mathematics was a matter of life and death. With the threat of the Luftwaffe overhead, the mathematician paced the frozen expanse of Lake Ladoga and inspected a caravan of precisely weighted trucks that were to trek over the ice.[1] The German and Finnish armies had besieged Kantorovich's hometown of Leningrad during Operation Barbarossa in 1941, and they had severed all the city's inbound roads and railways to starve it into submission. Yet, a single thread tied the city to the outside world, one that took the unlikely path from its eastern flank over Lake Ladoga. While Soviet barges could ferry supplies in the summer, in the winter sleds and trucks made the perilous journey on ice. This 'Road of Life' was the only way to keep the millions of trapped civilians and soldiers in Leningrad alive and fighting. There was much death on this road too – some forty trucks fell through the ice during the first week of the winter convoy.[2] Kantorovich's job was to minimize these losses. If he failed, the city would not hold out for long.

His task was to solve an urgent mathematical problem: given the wind, temperature, and thickness of the ice, how many trucks could be sent over the lake, and how heavy could they be? Rapid changes in weather conditions and the threat of German planes made the puzzle even more difficult. Despite the danger, the young professor insisted on being on the ice himself to see the convoy through these challenges. Kantorovich's efforts brought thousands of tonnes of fuel, food, and munitions into the city, and nearly 1.5 million civilians out of it. Adolf Hitler thought he could conquer Leningrad in six weeks; nearly 900 days later, the siege was lifted, and the humbled Wehrmacht retreated westwards.[3]

When he was not busy calculating the margins of life and death at Lake Ladoga, Kantorovich was hard at work on his masterpiece, *The Best Use of Economic Resources*. While his early mathematical work had been in the abstract fields of analysis and topology, this book was as practical as the Road of Life. Kantorovich's study delineated how 'mathematical methods' could be applied to the 'whole economy' on a 'scientifically planned basis'.[4] While he stressed that a capitalist economy could never approach such a degree of rationality, he politely recognized that 'planning deficiencies exist as a direct result of economic science lagging behind the requirements needed in the building up of a communist state.'[5] *The Best Use of Economic Resources* was an attempt to provide an economic science commensurate with the utopian ambitions of the Soviet Union. In place of the self-interested and often inefficient decisions made by the central planning bureau, Gosplan, Kantorovich imagined that algorithmic planning could increase efficiency at every scale, from factory to nation. Just as he had optimized the convoys across the icy lake, the young mathematician sought to optimize socialism itself.

The unlikely birthplace of this dream of 'red plenty' was a plywood factory in 1938.[6] As Soviet science emphasized applied work for ideological and practical reasons, it was natural enough that the plywood factory's engineers presented Kantorovich with the problem that would define his career: optimize production at a factory with eight different lathes, each with different speeds and outputs depending on the type of material processed, and five plywood types to be produced in a specific ratio. The problem was much more difficult than it appeared. Kantorovich soon realized that if he were to use traditional methods of optimization, he would need to solve about 1 million equations. He raised the matter with his colleagues in the department, but they had no solutions – apparently the engineers had visited the mathematicians before, to no avail.

Kantorovich pondered the problem all summer, and soon 'engineering and economic situations started to come into my head' that he realized were also problems of constrained maximization.[7] Within a few months he had his solution. He first presented his findings that October at the Herzen Institute, and in 1939 he published the path-breaking article 'Mathematical Methods of Organizing and Planning Production'.[8] He called his algorithm 'linear programming', and it allowed him – in an era before electronic computers – to work out a factory's optimal arrangement in an afternoon using only pen and paper.[9] What's more, the algorithm could be universally applied to any situation where a particular value subject to linear constraints had to be maximized or minimized. Linear programming was not only a quintessentially socialist kind of mathematics, 'characterized by a constant overlap of theory and practice', it also offered a new kind of socialist political economy.[10] Kantorovich immediately began to imagine how his method could be scaled up.

Although Kantorovich seemed unaware of Neurathian in natura calculation, linear programming offered what was perhaps the first practical method to institute moneyless planning. Rather than reducing everything to a universal equivalent (like a price), Kantorovich could balance competing restrictions in their natural units – tonnes of steel and concrete or hours of labour – across many different projects simultaneously. While not sufficient on its own to plan something as complex as an economy (as we will discuss further in this chapter), linear programming was a conceptual breakthrough. As we saw in the first chapter, neoliberal economist Ludwig von Mises was sceptical that planners could efficiently distribute key intermediate goods, such as steel, without the aid of prices. Linear programming offered a systematic way to allocate resources, so that it optimized some metrics of overall national well-being (which we will explore later).[11] Suddenly, in natura planning seemed possible, with informational requirements modest enough to make Mises blush. That is, as soon as a planner could articulate the material constraints of an economy using mathematical language, plans of production and distribution could naturally follow without the aid of the market's invisible hand. Otto Neurath, who was still alive in 1939, was unfortunately unaware that there was now a rebuttal to the neoliberals' epistemological critique of socialism. Indeed, even with the primitive computers available in the 1940s, Kantorovich could dream of 'programming the USSR'.[12] From the very beginning, Kantorovich argued that his vision was 'connected specifically with the Soviet system of economy' because large-scale planning problems 'do not arise in the economy of a capitalist society'.[13]

Kantorovich's breakthrough did not earn him any laurels at first. Criticizing the inefficiency of Stalinist planning was unwise, to say the least. Naively, he submitted a report on linear programming to Gosplan in 1942 and presented it at the Moscow Institute of Economics that same year. Gosplan rejected the proposal (and did so again the following year), and his discussion with economists there was 'quite sharp'.[14] Even Kantorovich's milder reforms were met with suspicion: after he optimized railroad car production, critics did not praise him for cutting waste but accused him of sabotaging the scrap-metal supply.[15] Some Soviet economists lambasted Kantorovich for not drawing explicitly on Marx's theory of value. Ironically, this critique overlooked how Kantorovich's mathematics would have been at home in debates during the early 1920s when moneyless economics was *en vogue* (even Neurath had merited close study in those halcyon days of the USSR).[16] Some alleged that Kantorovich's framework resembled the neoclassicism of the 'fascist' economist Vilfredo Pareto, because both used advanced mathematics.[17] It was a small miracle that Kantorovich survived these dark years; indeed, he only learned later on that his needling of Gosplan had been 'dangerous'.[18] It is remarkable, then, that despite these setbacks and the intense anti-Semitism of the period, Kantorovich managed to win the prestigious Stalin Prize in 1949. This was because his powerful patrons in the Red Army wanted to reward his service in the atomic weapons programme.[19] Recognition for his breakthrough of linear programming took longer.

In 1956, the new general secretary, Nikita Khrushchev, denounced Stalin's crimes, sparking a 'thaw' in Soviet society. Kantorovich was welcomed back from the cold, and his career soon blossomed. In 1958, he became a corresponding member of the Soviet Academy of Sciences, and a year later he was finally allowed to publish *The Best Use of Economic Resources*. By 1961, forty institutes dedicated to the study of mathematical economics had sprung up in what had been a bleak intellectual landscape just a few years prior.[20]

In 1960, Kantorovich moved from Leningrad to Novosibirsk, which hardly sounds like a promotion, but for a time this Siberian town became the world's premier centre in planning theory. An unspoken attraction was the unusual degree of academic openness permitted in isolated Novosibirsk compared with the metropolitan centres of Moscow and Leningrad, inverting the Siberian experience of exile into one of freedom. This was especially true for minorities such as Jews, who faced hiring quotas elsewhere. Kantorovich was at the height of his influence during his decade in Novosibirsk, where he not only won the Lenin Prize (the Soviet equivalent of the Nobel) but also served as the deputy director of the Central Economic Mathematical Institute (CEMI). CEMI housed economists and mathematicians in a building designed to be half for humans, half for computers – the machines were meant to be housed on double-height floors.[21] In many ways, Kantorovich embodied the optimism of the period, where rapid economic growth, the new universal science of 'cybernetics', and the space age seemed to herald the coming of an abundant and humane socialism.[22] Yet despite these seemingly promising conditions, Kantorovich's ideas were never implemented on a national scale.

There were two reasons for this. The first and more proximate cause was the Prague Spring in 1968, when a reformist 'socialism with a human face' spooked party elites of other Warsaw Pact states, who then invaded Czechoslovakia. After this crisis, anything that smacked of 'market socialism' (a tradition that Kantorovich arguably did not belong to) was compromised.[23] The episode weakened the Soviet prime minister, Alexei Kosygin, who was personally interested in new methods – like cybernetics – and who sponsored reform initiatives such as Kantorovich's base in Moscow, the Institute of Economic Management. The crisis empowered Kosygin's conservative co-consul, Leonid Brezhnev, leaving reformers little chance of revitalizing an increasingly decrepit state apparatus.[24]

The second reason for linear programming's failure is more profound, lying beneath the surface phenomena of academic squabbles and short-term crises: the lack of democracy in the Soviet Union. This meant that it was impossible to assemble a new political coalition strong enough to overcome the vested interests of economic planners and managers, who enforced the Communist Party's series of five-year plans.[25] Optimizing the whole economy would rob this elite of their power over the distribution of resources. Thus, despite the efflorescence of Soviet planning theory in the decade before 1968, Gosplan remained just as hostile to real change as it had been in 1942. Indeed, it was easier for Soviet reformers to get information on the national economy from CIA reports than from their own country's State Committee for Statistics.[26] Reformers were allowed to optimize various factories or even industries, but never the economy as a whole. Even a highly decorated technocrat like Kantorovich could not realize the dream of an efficient, moneyless socialist economy because no social movement existed that could help him overcome elite opposition.[27]

Conscious control is a planned economy's greatest strength, but it requires democracy to prevent authoritarian and inefficient control over the production and distribution of goods. In this way, we could say that linear programming remained a doomed demi-utopian proposal as long as it was not integrated within a broader political project. Democracy has become only more necessary in the globalized world that socialism will inherit, in which different locations will have specialized roles in the economy and will require supplies produced in faraway regions. An extraordinary co-ordination effort will be required to ensure that no one is left out or exploited in this global network of interdependence. As Kantorovich understood it, the goal is not to micromanage every kilogram of coffee or piece of steel rebar around the globe, but to 'construct a system of information, accounting, economic indices, and stimuli which permit local decision-making organs to evaluate the advantage of their decisions from the point of view of the whole economy'.[28]

Thus, it is necessary to marry Kantorovich's technical vision to Neurath's democratic socialism, in which planners lay out their goals and constraints in natural units and then devise different plans that could be chosen by an informed public.[29] These plans would represent many possible futures for a socialist planet – one might involve geoengineering and the conveniences of fossil fuels, while another could abolish the use of hydrocarbons entirely. The costs of each of these possible futures can be estimated in natural units, making clear the difficult trade-offs that must be made. Parliamentary representatives could decide on a plan, or perhaps the choice could be put directly to the people in a referendum. Creating plans based on in natura calculation and putting them to a vote would demystify the economy, making it more difficult for a selfish bureaucratic caste to obscure and thus control its workings. Although the methods required to co-ordinate today's economy would be considerably more complex than the linear programming Kantorovich originally employed, the need for democracy is no less pressing.

In this chapter, we will detail how to organize a democratically planned economy in an age of ecological crisis. It does not follow that such a task would be easy or that a socialist economy would produce limitless abundance. Indeed, ecological considerations make high rates of economic growth in perpetuity an impossibility, and therefore a constrained 'steady-state economy' – an economy that maintains a constant size, without increasing the throughput of natural resources – is a more likely condition at the end of history. Even a wisely managed eco-socialist utopia would still be plagued by some inefficiencies and shortages, as we will discuss later in the chapter. However, we believe that it is worth paying to gain other advantages, such as a stable climate, wondrous biodiversity, and a respite from pandemics. Half-Earth socialism also promises the prospect of a unified humanity, peace, and equality, with an economy built around care, health, and unalienated labour. Today, capitalism has never offered a bleaker future, while socialism has never been more feasible and necessary.

Crossing the Ice

Creating a just world that fits within ecological constraints is the Road of Life that humanity must cross in the twenty-first century. During the siege of Leningrad, Kantorovich understood that trucks would crash through the ice if they were loaded too heavily, but if they were loaded too lightly, more people would freeze or starve. Half-Earth socialism requires a similar balancing act, supplying everyone with the material foundations for a good life – sustenance, shelter, education, art, health – while protecting the biosphere from destabilization. In the scientific literature, this challenge is known as the 'planetary boundaries' debate, in which scientists calculate how to support everyone's basic needs without trashing the planet.[30] Such a research programme, however, is incomplete if it fails to recognize the impossibility of reaching these goals within capitalism. As we emphasized in the first chapter, the need to plan and constrain humanity's interchange with nature conflicts with the unconscious and expanding force of capital.

Although there are many estimates of planetary limits, even the most advanced models are unable to help us imagine post-capitalist ecological stability. This is not for a lack of technical know-how. Systems engineers at key institutions have built massive supercomputer programmes called 'Integrated Assessment Models' (IAMs) which combine physics, chemistry, biology, and economics into a single simulation of the world for the next 300 or so years. The IAMs used by the IPCC are usually a combination of a 'global general equilibrium model' (i.e., supply and demand are balanced across different markets) and a simulation of the climate, biosphere, and other natural systems. For example, the model might calculate global demand for energy, the resulting pollution created by that demand, and the effect those emissions have on the economy.[31] IAMs are central to climate politics. Anytime one hears a prediction about climate projections for 2100, an engineer with an IAM has probably been tinkering behind the scenes with variables such as pollution taxes, probability of technological breakthrough, spatial patterns of agriculture and biofuels, global food demand, the makeup of energy systems, and the sensitivity of the climate and biosphere to all these social changes.[32]

However, while there is much to admire in these ambitious simulations, IAMs are a clear demonstration of Neurathian 'pseudorationality'. For instance, BECCS is favoured by IAMs not because it is an effective or realistic solution to climate change, but because IAMs rely on the universal equivalent of money (even transforming CO_2 into cash via a carbon tax), and BECCS is a useful way to transform dollars into negative emissions *within the model*. Give a BECCS plantation x dollars a year, likely from a carbon tax, and you will sequester y kilograms of carbon from the atmosphere. Pseudorationality has now given the illusion that climate change can be reduced to a simple algebra problem. Clearly, another kind of global model – one linked to eco-socialism – is needed. This method should allow us to think in terms of trade-offs between discrete and incommensurate goals, much like Kantorovich's calculations on Lake Ladoga, without money or other universal equivalents distorting our plans.

This is not to say that the modellers themselves are clueless – indeed, many systems engineers understand that an enormous revolution in energy systems, drastic cuts to individual consumption, and radical resource redistribution from Global North to South are required to create a just society with a stable biosphere. Some have used IAMs to study the effects of truly staggering reductions of consumption and energy use in the developed world, paired with significant increases in the standard of living in the developing world – changes which would take us a long way towards the delicate balance demanded by our ecological Road of Life.[33] However, when modellers try to create radical IAMs, they usually just reveal the inadequacies of the global models on offer today. Like the scientists who research planetary boundaries, too many modellers lack a political programme that is able to realize the transformation they dream about. In this way, the position of radical IAM engineers is not unlike Kantorovich's at the height of his influence in the 1960s; prestige and knowledge count for little when stymied by powerful vested interests.

However, the situation is far from hopeless. After all, radical science combined with a large social movement can be a powerful force, like the anti-nuclear movement of the 1960s. Nothing scares neoliberals more than radical science allied with social movements, but until such a union arises, they have little to fear. Without a drastic political rupture, modellers will continue to be forced to rely on ever more unlikely deus ex machina, such as BECCS and SRM. Although critics of the Left often accuse socialists of magical thinking, the real fantasy is modelling a future where capitalism can be constrained within planetary boundaries.

Scientific Utopianism

Rather than IAMs rooted in the status quo, we need a radical global modelling predicated on what Neurath called 'scientific utopianism'. For him, 'utopia' did not refer to 'impossible happenings' but a 'thoughtful order of life' that did not yet exist. Whereas utopians from Thomas More to Edward Bellamy had pursued 'dreamers'' work – a worthy task – Neurath believed that by the early twentieth century such theorizing necessarily transformed into 'scientific work preparing the shaping of the future'. Utopias were not so different from the 'constructions of engineers', akin to detailed blueprints for an imagined society, and thus the utopians could justifiably be called 'social engineers'.[34] These social engineers would have to know about everything from the 'psychological qualities of men, to their love of novelty, their ambition, attachment to tradition, willfulness, [and] stupidity', as well as the 'natural base, land and sea, raw materials and climate' upon which society depends.[35] A utopian's aims might even include 'nonhuman ideals'.[36] Although what Neurath had in mind included 'the greatness of God' and 'the nation', expanding the conception of the good beyond humanity allows planning to incorporate ecological goals.

While Half-Earth socialism is a scientific utopian project, that does not mean that we are restricted to the tools and concepts that Neurath used. Even Kantorovich's sophisticated mathematics is hardly cutting edge now. While the static, one-off calculations made using linear programming are a valuable tool in managing any complicated project – the method is ubiquitous in contemporary applied mathematics, including in planning renewable energy systems – we will need other tools to allow local administrators to meet the needs of the people they serve, while simultaneously achieving global goals such as rewilding or long-distance trade.[37] Over the course of this chapter, we will consider various historic approaches to planning and modelling, and use them to devise a set of methods that will allow us to construct our own Road of Life.

Let's start with understanding in natura calculation. It does not mean replacing money with an inefficient barter economy (x kilowatt-hours of power equals y bushels of grain) but rather with an information system that sees how different goods relate to one another as a whole. Meeting the needs of nature and humanity is fundamentally a material goal, measured in food and carbon molecules, and seeing the world in natural units allows us to directly confront trade-offs without the obfuscation of money. While we have learned from a long line of previously proposed moneyless planning schemes, such as the one outlined in Paul Cockshott and Allin Cottrell's 1993 book Towards a New Socialism, many of these plans rely excessively on Marx's concept of 'labour time' to organize production and distribution.[38] The Cockshott and Cottrell plan, no outlier in this regard, rewards workers with vouchers representing the amount of labour they have performed, which can then be exchanged for goods that embody a similar amount of labour. Such schemes miss the forest for the trees, expending enormous effort in designing a system that corrects for the inevitable distortions created by labour money (how to account for the fact that some people are more effective workers than others, or that some kinds of work are more difficult or skilled than others, or that some items will be demanded more or less than their labour price suggests), when the goal of socialism is to allow humanity to consciously regulate itself and its interchange with nature.

Neurath argued that plans based on labour time are as pseudorational as capitalist profit, in that both are based on a universal equivalent that obscured more than it clarified. This is why Neurath portrayed socialist democracy as choosing one among several competing 'total plans' devised by the social engineers. Each total plan represents a distinct vision of how the productive capacity of society can be deployed. Neurath was never especially clear on how to devise such plans, but linear programming would be a powerful, albeit rather blunt, tool that would allow in natura planners to translate abstract goals into a concrete vision. At this point we can begin our work as social engineers, equipped only with the simple device of linear programming. As the chapter develops, the limits of this tool will become clear, which is why we also consider complementary methods.

Let a Hundred Gosplants Bloom

Imagine that the Half-Earth socialist revolution happens tomorrow. Before the new regime starts the hard work of planning the entire economy, they commission social engineers to do some quick calculations on balancing human needs with planetary boundaries and assign them to the new 'Gosplant' bureau (forgive us). Gosplant's initial goal is to devise several futures that illuminate how much of nature can or should be humanized to furnish the global economy with the necessary resources. It might model futures which allot higher or lower per-capita energy use, assume different rates of technological or infrastructural progress, and commit to varying degrees of rewilding, making visible the different obligations people will have to shoulder if they wish to achieve certain ecological goals. For even a simple version of this simulation to work, however, information must be gathered on a global scale and protocols established that can translate chaotic reality into natural units. We will first follow the Gosplant planners as they pull together the necessary information for their models. Then, we will run our own version of the linear programming algorithm to see the trade-offs Gosplant might confront.

The scientific corpus on planetary boundaries allows Gosplant planners to mathematically express two essential restrictions: limiting extraction to keep the biosphere healthy, while equitably distributing enough natural resources to supply human needs. As we saw in chapter two, it is necessary to set half the earth aside for rewilding to limit the ecocide of the Sixth Extinction. On top of this territorial restriction, scientists have provided global figures on myriad other ecological limits: from the maximum amount of nitrogen and phosphorus that can be used as fertilizer (62 and 6.2 megatonnes per year respectively) without causing mass eutrophication, to the freshwater available for consumption (4 petalitres per year), to the carbon that can be emitted (1.61 tonnes per person per year for 2°C of warming, even less for the more ambitious 1.5°C target).[39] Further restrictions on acceptable levels of pollution can be taken from the public health literature – for example, fine particulate matter suspended in the atmosphere should have an annual mean of 10 micrograms per cubic metre or less.[40] None of these limits should be taken as immutable – scientific knowledge reflects not only technique and theory but also social concerns. To what degree we humanize nature is ultimately a Neurathian decision with incommensurate ethical, technical, social, and political dimensions. It is a choice that we must consciously make and revise without recourse to any pseudorational universal metric.

Remember, the Gosplant planners have to figure out how to

not only stabilize the biosphere but also provide for everyone's needs. Energy and food production entail significant environmental costs that must be simulated in any global model. To plan for consumption in these two sectors, Gosplant will make several plans with different power quotas, like that proposed by the 2000-Watt Society we encountered in chapter two, and ensure that everyone in the world will have access to a nutritious diet. These choices, like many of Gosplant's other constraints, are all social decisions. Some scholars have ventured energy quotas lower than 500 watts, which is less than a twentieth of current US usage. This is not to say these more conservative targets are impossible – they would just require spartan restrictions on new clothing, appliances, transportation, electricity, and living space.[41] Gosplant need not make this choice for us; they can generate many plans, each with its own energy quota, and leave it to the people and their representatives to decide which plan best balances the needs of the biosphere and humanity.

Linear programming is a powerful protocol which can turn these various constraints (expressed in natural units) into concrete plans. We toss in limits on resource use and the minimum needs of humanity and turn the crank. The Gosplant social engineers need not harbour *any explicit preference* between diets, energy systems, and other variables. There is no inherent requirement in their linear programming model that the world be vegan, or that all power sources be renewable. Instead, in our toy example, the planners set their two main goals – providing enough food and energy for everyone's basic needs and staying within planetary boundaries – as well as the basic productive configuration needed to satisfy those goals by different means. The final piece of the puzzle is called the *objective function*: the quantity that the linear programming algorithm must maximize or minimize. A capitalist firm might decide to minimize costs when running a linear programming algorithm on its own operations; the Gosplant planners, on the other hand, might opt to minimize land use, carbon emissions, or some other metric that combines multiple goals. The linear programming model will then output the best mixture of energy and food sources with respect to this objective function, or tell the user that the plan is not possible within the given constraints.

To begin to understand how linear programming digests its inputs, consider the question of diet. Under Half-Earth socialism, everyone could eat a healthy diet, but there are multiple ways Gosplant could achieve this goal. If the great majority of people were omnivorous, this would impose a per capita cost of 1.08 hectares of land and 2.05 tonnes of carbon per year. Vegetarianism does much better, taking up only 0.14 hectares per person and 1.39 tonnes of carbon; from there, veganism pushes land use down slightly, to 0.13 hectares, and emissions more substantially, to 1.05 tonnes.[42] A slew of other reforms could – and indeed must – improve these emissions figures much more (land use would be harder to change significantly).[43] As you might have guessed, even though the Gosplant bureaucrats did not explicitly prefer any diet, their linear programming algorithm will likely opt for veganism because it satisfies the requirement of feeding everyone with the smallest environmental impact. If a social movement plays Glaucon to our Socrates, then the planners could include some meat production. However, increased emissions and agricultural land use will cut into other aspects of the total plan. Linear programming is only a tool, but it is one that allows the Neurathian politics of seeing and democratically deciding which trade-offs to accept.

Having just two metrics to measure the world – land use and carbon emissions – skirts close to pseudorationalism, though it is enough for a toy example. In reality, Gosplant planners would consider the future from a variety of angles (what Neurath called a 'silhouette'). Take veganism, for example. The transition to a plant-based world would impose greater sacrifices on the carnivorous Global North, which is only fair – the average North American eats almost ten times more meat than the average African.[44] The recent EAT-Lancet study proposes a nearly vegan 'planetary health diet' that allocates everyone a 2,500-calorie quota that would not only lessen humanity's impact on the environment but also prevent an estimated 11 million deaths per year. There would be less malnutrition, as well as fewer non-communicable conditions such as type 2 diabetes and heart disease caused by the over-consumption of meat and certain processed foods.[45] Replacing livestock fodder with pulses and legumes would increase natural nitrogen fixation (and thus reduce the need for a fossil-fuel-dependent fertilizer industry), while allowing pasture to be rewilded. A more fully developed linear programming model might account for these benefits in more granular detail, better reflecting the myriad social and ecological benefits of veganism. The ethical benefits of planetary veganism may be more intangible but still relevant to a Neurathian public debate. The world's moral issues will never be solved by a computer, but algorithmic planning can clarify the discussion.

The social engineers would address the energy question in a similar manner to food. Each person on Earth needs to be supplied a power quota, whether 2,000 watts or something else, but again there are many ways to do so. Suppose Gosplant has eight main sources of energy to choose from: photovoltaic solar cells, concentrated solar power stations, wind turbines, biofuels, nuclear, methane ('natural gas'), coal, and petroleum. For simplicity's sake, and because the anti-nuclear movement was a vital constituent in the Half-Earth socialist revolution, the planners do not include nuclear energy in their initial calculations.[46] Each of these power sources has a cost associated with it, expressed in natural units of land area and carbon emissions, but not money. Biofuels, for example, are hypothetically zero-carbon (though this is often not true) but have a low power density: around 0.23 W/m^2 for US corn-based ethanol, 0.5 W/m^2 for ethanol made from sugar cane, or 0.5 W/m^2 for woody phytomass.[47] Airplanes, reliant on efficient kerosene, would need stronger stuff, made from soybeans (a paltry 0.06 W/m^2) or that sordid simian slayer, palm oil (0.65 W/m^2).[48] Methane would emit 3.6 kg CO_2/W over a year but has an enviable power density of 4,500 W/m^2.[49]

Solar and wind are the best hope for renewables, with power density an order of magnitude greater than that of biofuels (8 W/m^2 on average across the US for solar and topping out at 50 W/m^2 for wind, though this can fall dramatically in large wind farms or stagnant regions) and none of the air pollution or water problems.[50] While a basic in natura analysis makes manifest the trade-off between CO_2 and land, a more sophisticated model could add other costs, such as the environmental and social costs of mining various materials. The idea remains simple: linear programming requires Gosplant only to lay out goals which have been democratically decided upon, as well as collect the information on the material costs of each variable.

Planning the energy sector is not merely a matter of tallying watts; it also takes into account the physical properties of various energy systems. The Gosplant modellers would have to reckon with renewables' other limitations such as location. Relying on natural flows rather than stocks of energy means there are relatively few places like sunny Andalusia that can host 'concentrated solar power' installations. Next, they need to think about how to handle the variability of renewable energy production – what happens when batteries run out and there isn't enough wind and sun to meet demand? Planners could opt to allow brown-outs or blackouts on such occasions, or use methane or biofuels as a 'back-stop' power source.[51] A linear programming algorithm could determine the ideal mix of biofuels and methane for such scenarios while still adhering to land-use and emissions constraints.

Energy use in transportation and industry is the next constraint that Gosplant has to set. Although much of the current debate over renewable energy focuses on electricity, that actually comprises only a small part of total energy use.[52] Currently, industry and transportation not only require more energy than the electrical sector but are also difficult to electrify.[53] For example, smelting iron requires high temperatures best supplied by coal or land-hungry charcoal. We are hopeful that 'green' hydrogen (which uses renewables to power the electrolysis of water) and expanding public transit would allow for the electrification of these challenging sectors.[54] Until then, the dreadful choice between fossil fuels and biofuels will remain necessary to some degree.

The long-term energy goals of Half-Earth socialism are apparent: total electrification of industry and transportation and making generous use of clean hydrogen where fuels remain necessary, with all power supplied by relatively power-dense wind and solar. Hydroelectricity, tidal energy, and geothermal energy may be useful, but they would always play a limited role in a renewable energy system. Ideally, such a society would not use any biofuels and would rely on batteries for back-stop power instead (though biodiesel back-up generators at hospitals and the like might be wise).

Half-Earth socialism will need to dedicate humanity's full productive capacity to realize this future as soon as possible, for this transition won't be easy. One study estimates that a renewable world would require 3.8 million 5-megawatt wind turbines (supplying 50 per cent of global electricity and covering 1.17 per cent of total land area); 1.7 billion 3-kilowatt rooftop solar panels; 89,000 300-megawatt solar plants, a little over half of which would use concentrated solar power, with the remainder using photovoltaic cells (all solar sources would make up 40 per cent of global electricity and take up 0.29 per cent of land area, not counting rooftop panels); and tidal, hydroelectric, and geothermal energy producing the remaining 10 per cent of electricity.[55] Unmentioned are the massive investments required for a hydrogen industry and public transit. The transformation would be extraordinary: supplying the necessary electricity would require a 40-fold increase in wind power and a 170-fold increase in photovoltaics as of 2015.[56] As Vaclav Smil points out, 'such a ramping-up of all kinds of capacities – design, permitting, financing, engineering, construction, all going up between one and five orders of magnitude in less than two decades – is far, far beyond anything that has been witnessed in more than a century of developing modern energy systems.'[57] Even an eco-socialist society that is single-mindedly committed to overcoming the energy challenge would struggle to pull off such a change.

Rolling the DICE

After calculating the land and emissions costs of various diets and energy systems, we now have enough data to run a simplified linear programming model like the Gosplant bureaucrats might devise when the post-revolutionary government requests a tentative global plan. Like some early IAMs, such as William Nordhaus' Bank of Sweden Prize–winning DICE model, our programme is basic enough to run on an ordinary laptop in less than a second. We can set several constraints, including an energy quota; various planetary boundaries, such as global temperature or the amount of land set aside for wildlife; and the state of infrastructure and industry (e.g., to what extent are they electrified?). Full details of the toy model are given in the appendix at the back of the book. While the dream of full electrification is apparent in the long term, if the revolution happened tomorrow, then Gosplant would have to decide whether to minimize land use or CO_2 emissions, in addition to setting energy and food quotas. The numbers underpinning these variables are based on the current state of technology in various fields – so no cold fusion or fast-breeders. The only futuristic element in our toy model is population, which we set at 10 billion people. This number, some 2 billion more than now, is the estimated global population in 2050. That Half-Earth socialism could provide a good life for our abundant species and still protect the environment makes clear that the Malthusian fear of 'overpopulation' is dangerously exaggerated.

Using this model, the Gosplant planners develop their first plan to tease out where the difficulties for a transition might lie. Pessimistically, they assume that transportation and industry remain unelectrified and consume the bulk of energy, as is the case in the US today, and thus require vast quantities of fossil fuels or biofuels to meet humanity's needs. They have three main goals: provide a 2,000-watt quota for all, limit warming to 2°C, and rewild half the planet. With their goals settled, all the Gosplant planners need to do is pick an objective function. They decide to minimize land use. However, when the bureau runs its model, it finds that the plan's goals cannot be met, even if everyone became vegan. As biofuels would compose such a large share of the global energy budget, there would be no way to grow enough food and energy crops without transgressing the Half-Earth threshold. This would then entail SRM or biodiversity loss caused by the huge biofuel plantations. If they wanted to, Gosplant could add these terrible possibilities into the model and relax their planetary boundary constraints. Although things seem dire, it is too early to give up on utopia!

The planners have several options. One would be to reduce the energy quota to 1,500 watts, which would make the rest of the plan viable even without electrifying transit and industry. According to the toy model, 57 per cent of the planet's habitable surface could be left to nature (up from 15 per cent now), 26 per cent would be dedicated to biofuels (up from about 0.4 per cent now), and 18 per cent to agriculture (down from 50 per cent now).[58] The assumption that allows this plan to work is that virtually everyone would be vegan. The model also shows that because energy use would be so low, methane could be used for some industrial processes and electrical generation while still limiting warming to 2°C. Although the plan is feasible, the planners are reluctant to create such a massive biofuel industry. They come up with another option by simulating strict restrictions on private car ownership and unnecessary industrial processes, which halves the demand for solid and liquid fuels. In this modified plan, biofuel crops take up only 21 per cent of the planet's surface. An even more ambitious option reduces the energy quota to 1,000 watts and would require biofuel plantations on only 13 per cent of Earth's surface, while reserving an astounding 70 per cent for wildlife.

Wonderful news! Even with quite pessimistic assumptions, Gosplant can plot several paths towards an equal, sustainable planet. However, the blueprint drafting cannot stop there. Anticipating possible demands from climate activists, the Gosplant planners devise another blueprint that opts for the courageous goal of limiting warming to only 1.5°C. After re-running the algorithm with the 1,500-watt quota and the restricted fuel-use scenario, their model shows that this goal will require the biofuel sector to expand to over 25 per cent of the planet's surface (up from 21 per cent before). Warming would stay below 1.5°C, but at the cost of having taken more land from nature due to much stricter restrictions on fossil fuels. There are no easy solutions here, and our Gosplant model clarifies the trade-offs required in every plan. Ultimately, a global parliament would have to take a vote on whether minimizing climate change or preserving habitat is the more urgent planetary goal – or whether the planners should go back to the drawing board and come up with more arrangements.

Other options become possible with new infrastructure and technology. Perhaps there is a breakthrough in 'green' hydrogen fuels, which allows the Gosplant social engineers to pursue the goal of total electrification. This leads to their most ambitious plan yet: an energy quota of 2,000 watts and 50 per cent of land rewilded, all within the limit of 1.5°C warming. After running the algorithm, they are thrilled to see that it works with plenty of room to spare! Electrification allows Gosplant to take full advantage of solar and wind power, which have much higher power densities than biofuels; with land use minimized, a whopping 81 per cent of land can be left to nature (thus preserving 95 per cent of all species according to Wilson's formula). The planners find that up to 24 per cent of the population could be omnivores in this scenario, as land constraints are so relaxed as to permit the return of some animal husbandry. Of course, a vibrant animal-rights movement would still oppose this for ethical reasons, while epidemiologists might warn against the threat of zoonotic disease. The point, however, is that the social engineers' plans could and would evolve alongside infrastructural and political change.

While full electrification would almost certainly be a long-term goal of Half-Earth socialism, we would have to confront certain challenges first. Imagine that, for the short term, the global parliament opts for the second modified plan with a 1,500-watt energy quota and restrictions on fuel use. It is the best fit for present circumstances while allowing energy use to grow in the future as more sustainable infrastructures are built. (That quota would seem austere in the Global North, though relatively painless for the South.) The government agrees to steadily reduce private car ownership to the point of complete abolition, one compacted Ferrari at a time. The steel thus saved can be recycled into trams and buses, while the remaining cars (which would run on electricity, hydrogen, or biofuels) are pooled and signed out by individuals or families. While Gosplant liquidates the suburban real estate market early on, millions of construction workers and tradespeople find work retrofitting buildings to conserve energy and adapting private mansions and corporate headquarters to communal use. Private lawns and golf courses are likewise either rewilded or turned into community gardens. Wide-ranging improvements to industrial processes to reduce pollution, fuel use, and waste are undertaken in just about every industry. Large swathes of manufacturing become rationalized when 'planned obsolescence' itself is made obsolete. Resources are redirected towards building solar panels, wind turbines, super-efficient insulation, and railways. Immediately after the Half-Earth socialist revolution, much of the world's pasture is converted into biofuel plantations for the short-term decarbonization of transportation and industry, while the remainder is rewilded, which in turn requires an expanded cadre of ecologists and foresters trained in both conventional science and traditional Indigenous knowledge.

Gosplant's job is not to dictate what the future should look like but to supply the public and its representatives with blueprints. For Gosplant, the process is more important than the final product. Linear programming offers a quick way to see what goals are possible under different scenarios, without relying on the market or market-dependent IAMs. Just as socialism is the ability to perceive and thus consciously control the economy, thought experiments like Gosplant allow us to imagine what socialism might look like in practice.

In Natura Democracy

Despite his espousal of 'total plans' to co-ordinate society, Neurath was deeply sceptical of technocracy. 'People of the totalitarian kind may try to make scientists the leaders of a new society', he warned, 'like the magicians, nobles, or churchmen of former societies.'[59] Neurath innovated new methods of communication and education to ward off such a threat. He believed that even something as complex as the economy could be made comprehensible to everyone, and that this was indeed the precondition for socialist democracy. To solve the riddle of economic perception, Neurath turned to art instead of to algorithmic planning.

Neurath wore many hats – that of philosopher, economic historian, planner – but he saw his role as curator as essential to inculcating class consciousness. During World War I, he not only worked as a military planner in Vienna but also directed Leipzig's Museum of War Economy, which explained the war effort to the German public. He was the head planner of the post-war Bavarian Soviet Republic, an entity destroyed in the spring of 1919 by the centre-Left SPD government in Berlin and its Freikorps henchmen. Lucky to escape with his life, Neurath returned to Vienna, and to curation. As an influential member of the city's allotment gardening (*Siedlung* or 'settlement') movement, he set up and ran the city's Settlement Museum, expanding its purview in the mid-1920s to become the Social and Economic Museum.

Instead of giving pride of place to rare objects, Neurath's museum exhibitions were almost entirely composed of charts and graphs illustrating various economic indices. To this end, he devised ISOTYPE (International System Of TYpographic Picture Education), a graphic language designed to convey complex economic information not only to Austria's working class but even to people who could not read German (or read at all). ISOTYPE was the fruit of his collaboration with Marie Reidemeister (who would become his third wife) and artists Erwin Bernath and Gerd Arntz. Inspired by Egyptian hieroglyphics, Neurath developed ISOTYPE as a pictographic language based on a few simple principles, including the ideas that individual 'symbols must be self-evident, clear in themselves', and that their combination 'may form a unit of information like a story'.[60] For example, ISOTYPE designers could show change over time by depicting farmer symbols transforming into factory worker symbols, with the number of each depicting the size of the workforce in agriculture and manufacturing. Rather than relying on a Platonic elite of logicians, Neurath thought that a visual language could democratize reason by making the essence of an economic problem apparent to non-experts. ISOTYPE, 'an education in clear thought', was designed to help teach the working class to see the economy, all the better to appropriate it for themselves. In this way, Neurath's approach to socialism was the mirror opposite of the neoliberals' mystification of the market.[61] Our own humble attempt to follow Neurath's example is a Half-Earth socialist planning game, available at http://half.earth. Yes, you, too, dear reader, can become a social engineer and devise your own scientific utopia.

Gosplant would similarly seek to educate citizens on how the economy and biosphere function, to create the possibility of democratic planning and ecological stability. While ISOTYPE principles are still quite useful to convey information, we can go a step further than Neurath, thanks to linear programming. Kantorovich's optimization model was brilliant even though the math he used is not particularly advanced. This means that linear programming and other planning tools could be a part of basic education, giving citizens the opportunity to pore over Gosplant's plans and understand for themselves how the world works. Half-Earth socialism would not be some distant, top-down technocracy but rather a relatively simple democratic system, based on robust public education and involvement. An informed citizenry would be well equipped to choose among the competing plans devised by the planners. Indeed, it would not be hard for citizens to draft their own rough blueprints, which could also be put to a vote. Such clarity is empowering in two senses: first, because it allows everyone to be a part-time planner and participate meaningfully in political discussions; and second, because one can see one's own work as a small but necessary component of the general intellect and labour of a unified humanity. This is what democratic control over production looks like.

Steering the Socialist Ship

Let's return to our imagined future, where the Gosplant planners are busily engaged in scientific-utopian speculation. As the world readies itself for the enormous transformations needed to create Half-Earth socialism, it becomes clear that Gosplant will be overwhelmed by the sheer amount of planning to be done. For all its pedagogical and democratic value, linear programming alone will not suffice to plan something as complex as the global economy. Co-ordination has to happen on all levels at once. The world's truck drivers and train conductors need to know where to deliver their cargo, at the same time that Gosplant has to calculate where to rewild the next 10 million hectares of former pasture. Not only will there be enormous variation in the spatial scales to co-ordinate, but the planners must also account for changes over time. It's not obvious how static figures computed using linear programming can handle these transitions, especially if projects fall behind schedule. Furthermore, how will the bureau deal with emergencies, such as a poor harvest or a devastating hurricane? Kantorovich's linear programming will not be enough in itself to create a global in natura economy. For a solution to this problem, Gosplant will need to supplement its simple linear programming simulations with insights from other disciplines, historical and contemporary. Its intellectual journey begins with a field that was fashionable in Kantorovich's time: cybernetics.

'Cybernetics', which comes from the Greek *kubernētēs*, or 'steersman', was coined by mathematician Norbert Wiener in his book *Cybernetics: Or Control and Communication in the Animal and the Machine* (1948). For Wiener, the heart of cybernetics is the problem of controlling systems characterized by what he calls 'feedback'.[62] In complex systems, such as a Gosplant global plan, the actions taken to control one part will in turn feed back and have their own effect on the system – for example, fully electrifying public transit reduces the need for biofuels and increases the need for solar panels and energy storage, freeing up more land for rewilding while simultaneously increasing pressure on the grid. Wiener developed the concept of feedback when tasked with shooting down German planes during the Battle of Britain in 1940. His 'anti-aircraft predictor' would take into account the defensive zigzag paths of enemy planes so that flak gunners could line up their shot.[63] The problem was not a simple one, since the anti-aircraft predictor had to take into account the turbulence of the air and a degree of randomness introduced by the pilots' actions. Wiener's breakthrough was conceptualizing that 'the pilot and gunner as servomechanisms within a single system was essential and irreducible.'[64] In mixing human and machine, the anti-aircraft predictor was the first self-consciously cybernetic system. Unfortunately for Wiener (and Britons living through the Blitz), he never got the system to work. Yet this unsuccessful research laid the foundation for the new field of cybernetics.

Soviet cybernetics was influenced by both Wiener's work and native intellectual traditions. The mathematician Andrey Kolmogorov, Wiener's Soviet counterpart, also researched probability theory in the 1930s, and between 1939 and 1941 he published a series of papers on the interpolation and extrapolation of stationary sequences, which Wiener himself admitted were 'equivalent' to his theoretical research on the anti-aircraft predictor.[65] Other Soviet cyberneticians, such as Anatoly Kitov (who worked on the Soviet space programme in the 1950s), read Wiener's prohibited *Cybernetics* in a classified library.[66] Like Wiener's own work, Soviet cybernetics was initially tied to the military, where its scientists built up their own theories of control through the development of complex technologies such as rocketry.[67] Similar to Kantorovich's concomitant mathematical research for the atomic bomb, Soviet cybernetics could *only* take place within the more pragmatic military sphere because these new fields were seen as a 'reactionary pseudoscience' by Stalinist scholars.[68] In 1958, cybernetics not only was rehabilitated but forced the spring of Soviet planning theory.

There were two main currents that shaped planning debates in the Soviet Union over the following decade: the theory of mathematical *optimization* (e.g., linear programming) and the cybernetic theory of *control*, built around differential equations.[69] *Half-Earth Socialism* is not a mathematics textbook, but it is worth describing the differences between optimization and control in some detail, as these two foundational methods in cybernetics have profoundly different implications for planning – and because this would be a crucial part of basic education under Half-Earth socialism.

We are already familiar with optimization, where one lays out an equation to be maximized or minimized, along with a set of constraints which limit the possible solutions. Gosplant social engineers could opt to minimize land use or carbon emissions, or to maximize the number of vegans – anything will do. Although optimization involves solving a single equation, the problem can be quite complex, including the optimization of multiple moments in time simultaneously (for example, optimizing biofuel production over a ten-year period).[70] However, the structure of optimization problems requires that any changes, however slight, be encoded in adjustments to the equation and constraints; from there the entire problem must be solved again. It's hardly ideal for the day-to-day challenges of Half-Earth socialism.

This is where control comes into play. Using differential equations, one can see how a system changes from one moment in time to the next by computing the *difference* between the present and the immediate next moment. For example, the speed of a car can be used to write a simple differential equation that describes its position – if it is moving at 30 metres per second, then in one second it will be 30 metres down the road, and in ten seconds it will be 300 metres away. Complex systems can often be modelled using differential equations, and control theory is used to gently adjust their behaviour so that the change over time goes as planned. Controlling a system usually involves a constant stream of input data, which is used to constantly correct the system in real time. Control theory can be thought of as our fictional car's cruise-control system, which monitors and adjusts speed many times as it takes in data from the wheels and engine to maintain a consistent outcome or speed. The result is a model that is continuously refined through feedback (e.g., information from the chassis) and has built-in flexibility to plan for change and contingency over time (say, a shifting slope), all in service of a pre-defined outcome (the speed limit). Perhaps you can already see how useful this would be for planning Half-Earth socialism.

In the Soviet Union, the debate over whether to emphasize control or optimization in economic planning was not so much about mathematics as it was about state power. Kantorovich disagreed with the control theorists because he believed optimization better fostered local autonomy.[71] Neither optimization nor control creates or inhibits freedom on its own – they are merely tools – but in the Soviet context Kantorovich was right to be suspicious. The cyberneticians were, like Wiener, based primarily in the military and thought about the economy as a machine to control, while mathematical economists like Kantorovich favoured optimization (with some caveats). Kantorovich understood that programming the USSR in one shot was a hopeless task and criticized a group of rival economists who were behind a programme called the System for the Optimal Functioning of the Economy (SOFE), which he believed relied too heavily on centralized optimization.[72] Instead, Kantorovich realized that optimization had to occur on multiple scales, both in space and time, with a vast number of models loosely connected to one another. Regional governments would have leeway over how the local economy operated, so long as it met the broad conditions set by the national plan. 'It is neither possible nor necessary to attempt to describe in detail and build all at once a final, all-encompassing system of models of the national economy', he reflected. 'It must obviously originate on the basis of a complex of individual models and tasks.' The goal was not a single model, but a 'single complex of interconnected models' designed with enough autonomy to 'provide working people and managers of individual levels and sectors of the economy broad opportunities to display their initiative.'[73] This multilayered planning is perfect for working within a natural and social world with enormous variation in ecology, climate, and needs in different areas.

Within the Soviet context, the optimizing economists envisioned a system that was far more decentralized than that of the control theorists. As historian Adam Leeds puts it, for cyberneticians 'the problem of planning became one of creating sufficiently pervasive, reticulated, and high-bandwidth information channels and sufficient computational capacity that the fused state-economy could essentially be a well-controlled dynamical man-machine system, akin to the antimissile systems', with little room for local variation.[74] Kantorovich, on the other hand, envisioned factories and farms independently using the parameters from the linear programming algorithm to plan their production, not unlike a price system but without a universal equivalent guiding economic decisions.[75]

Kantorovich's vision of multilayered planning is significant for several reasons. Firstly, it makes planning more computationally feasible. Even today's computers would struggle to optimize all the variables involved in a system as complex as the global economy.[76] This is related to the second point, which is that Kantorovich's multilevel approach reduces the amount of information needed to make the plan work. Hayek's argument that a planned economy requires 'one single mind possessing all the information' was either naive or cynical, but it is not a convincing rebuttal to Kantorovich's vision of socialism.[77] Thirdly, and perhaps more importantly, these loosely connected models create space for democracy, diversity, and political initiative, all unified in pursuit of national or even global goals. Such an approach is superior to what tends to be on offer in socialist political theory: either tiny but loosely connected anarchist communes or a top-down authoritarian structure.

How could Kantorovich's multilayered planning system be realized? It is not clear how information could flow between these many planning offices, nor how the optimal plan could be updated in real time as information flowed in from the real world. Not unlike our imaginary Gosplant, this system would have a hard time responding to shocks, and we can certainly expect many of those in an era of environmental crisis. To solve the problem, we will need to fuse Kantorovich's optimization with the cyberneticians' control. However, the Soviet control theorists do not offer an approach flexible enough for a democratic eco-socialism – ideally, our control system will allow for local variation, as Kantorovich envisioned, rather than creating a mathy military dictatorship.

Hold My Beer

It is at this point that we turn to another planning theorist to help us imagine our scientific utopia – Stafford Beer. Like Wiener, Beer was keenly read by his Soviet peers, but Beer was an unusual cybernetician in his deep commitment to democratic control systems.[78] He was not a typical socialist but looked the part of a 1970s business hippie – being a management consultant with a beard of biblical proportions – but he was far more like Neurath and Kantorovich than like Stewart Brand or Ed Bass. Beer believed that controlling 'exceedingly complex systems ... indescribable in detail' was the central problem of the mid-twentieth century.[79] This might sound Hayekian, but Beer distinguished himself from the neoliberals in believing that such systems *could* be controlled – even something as complicated as the economy – so long as the controller was itself complex enough to fairly represent the system. In cybernetics, this is known as the law of requisite variety.[80]

Acting on this intuition, Beer developed a management architecture called the viable system model in the early 1970s. The details are complicated, but the basic principle is that planning could be performed by a five-part, loosely hierarchical system.[81] As historian Eden Medina describes it, the first level of the system is the 'sensory level', similar to the parts of the body in contact with the outside environment.[82] Just as lungs breathe without conscious intervention, nodes on the first level usually operate without much input from the rest of the system. Individual groups of workers are capable of managing their own affairs, as the heart, lungs, and liver do. The second level is a support system for level one, which Medina calls the 'cybernetic spinal cord', connecting the different nodes on level one together in communication. The main purpose of the second level is to help organize level one's activity, allowing the different nodes to 'coordinate their actions and adapt to one another's behavior', as well as filtering out important information to send up to level three.[83] This third level was likened by Beer to the parts of the brain that regulate basic bodily function, the brainstem and cerebellum, in that it manages the operation of the first two levels. The third level has access to a detailed picture of daily activity and can thus co-ordinate the actions of the nodes on the first level, intervene when necessary, and send important information up to the fourth and fifth levels of the system.

The fourth level is where Gosplant would live, as it is where medium- and long-term plans are devised. As Medina notes, because daily management is affected by plans with a longer horizon, 'Beer did not see System Four as the boss of System Three but rather as its partner in an ongoing conversation.'[84] The highest level unifies the entire system and consciously intervenes in emergencies. Perhaps most importantly for us,

Beer's model exemplifies the promise of using control theory to organize an economy at multiple levels, without relying solely on optimization, and avoiding Soviet cybernetics' tendency towards rigidity. We should note that, unlike systems we will encounter later in this chapter, the viable system model is not a mathematical model but a management one which, inspired by the principles of control, makes liberal use of technology to aid implementation (e.g., in the design of the communication system on level two).

Unlike Kantorovich (but like Neurath), Beer had the opportunity to put his ideas into practice at scale. Beer's apotheosis from mere theorist to socialist immortal occurred not because of his influence on Gosplan or CEMI but through his participation in the ill-fated utopian 'Project Cybersyn' (or 'Synco' in Spanish). Although Chile in the early 1970s might seem far removed from the global crises in the 2020s, Cybersyn is perhaps the closest historical analogue to Half-Earth socialism. Beer designed the system to manage Chile's state-owned industries as part of President Salvador Allende's peaceful transition to socialism. The task was challenging, in part because the number of state-controlled factories, mines, and infrastructures increased by the day, and each required cutting-edge management and technical expertise. The viable system model was a natural choice for this fiendishly complex problem; like Kantorovich's meeting with the plywood engineers, this was a moment when theory and practice joined to further socialism. New technologies, such as telex machines, promised to overcome the problem of co-ordinating information in a far-flung country.[85]

Although Beer was not a Neurathian, much of his viable system model can be reconciled with in natura economics. This is because Beer's model was designed to manage any complex system, whether the internal operations of a company or the co-ordination of an entire economy, and therefore could be based on natural units rather than money. However, the similarities go deeper. Neurath's writings are filled with charts that look like the present-day 'block diagrams' used to design control systems. In one essay, Neurath drew links from the highest level of planning, his proposed Centre for Calculation in Kind (which was not unlike the top two levels of Beer's viable system model), to panels of experts and local planning offices (like the co-ordinating efforts of level two), and then down all the way to individual factories (the nodes on level one).[86] Just as Kantorovich showed how Neurath's in natura economics might work in practice, Beer's model demonstrated a practical way of implementing Kantorovich's dream of multilevel planning. However, Beer's model still used money as a universal equivalent, and significant parts of the Chilean economy remained in the private sector. Nevertheless, Neurath would have found much to admire in Beer's design. The viable system model allowed for both 'requisite variety' that could respond to local conditions and a central plan to guide investment.

Beer's multilevel design proved its mettle in October 1972, when the CIA organized a general strike of the bourgeoisie. Thousands of retailers, doctors, lawyers, engineers, factory owners, and private bus operators joined in, but most dangerous of all were the owners of freight companies and their 40,000 truckers.[87] Cybersyn allowed the government to survive this reactionary onslaught. The telex office served as the administration's war room, since it allowed key officials to communicate directly with workers (including the remaining 200 loyal truckers) and co-ordinate their activity. Industries could report shortages, and the command centre could locate a truck, identify an unblocked route, and deliver the necessary supplies. This was the viable system model in action: workers kept their factories running and even created new machine shops to repair the government's improvised shipping fleet, while officials in the telex room directed resources and collected vital information. Some factories even distributed goods directly to workers, bypassing the private sector entirely. As

Medina puts it, 'the network offered a communications infrastructure to link the revolution from above, led by Allende, to the revolution from below, led by Chilean workers and members of grassroots organizations, and helped co-ordinate the activities of both in a time of crisis.'[88]

Not all of Beer's projects worked at the time, but they did prove fruitful in the long term. The CHECO (CHilean ECOnomy) subproject was an attempt to create an 'experimental laboratory' for engineers to test the impact of policies in an economic simulation based on neoclassical principles, not unlike what IAM engineers do today.[89] However, as Medina argues, CHECO's reliance on these assumptions proved a grave mistake. For example, CHECO calculated inflation according to 'the levels of goods and services, productive capital, available capital, investment funds, prices, and total currency', variables which are reasonably predictive in relatively stable economies like those of Britain or the US, where these theories were developed.[90] However, CHECO was of little use when the US Export-Import Bank downgraded Chile's credit rating to junk in 1970 and blocked Santiago's access to foreign credit to further weaken Allende's government, which was already battling shortages caused by hoarding and capital strikes. With all the extraordinary events of open class warfare violating every neoclassical assumption, Medina asks, 'even if members of the CHECO team had somehow been able to identify the extent of U.S. meddling in Chile's economy, how could they have modeled it?'[91] In the end, the Chilean planners declared CHECO 'a failure'. Beer himself admitted that 'no one was anxious to place reliance on the results' because of the unreliability of the simulations.[92] While Cybersyn proved to be a powerful tool for co-ordinating conscious efforts on a nationwide scale, modelling based on the universal equivalent of money was less effective.

Beer never had a chance to improve on Cybersyn's design because the programme soon came to an abrupt and bloody end. After years of destabilizing the economy and pressuring the army to overthrow Allende's government, Washington finally got the coup it wanted in 1973. Without a Road of Life to help his government endure the fascist siege of the presidential palace, Allende shot himself. Beer was fortunate to leave Chile alive. One of the first acts of the newly installed junta was to destroy Cybersyn; a soldier even stabbed the transparencies in the central operations room.[93]

The new tyrant, Augusto Pinochet, created a marketized society protected by a strong, ruthless state. Death squads crushed the Left, while economists trained at the University of Chicago oversaw sweeping privatizations. While the 'Chicago boys' have become infamous in the history of neoliberalism, it is less well known that the CIA's goal of creating a 'coup climate' was largely the work of assistant secretary of defence Warren Nutter, who was Chicago School leader Milton Friedman's first graduate student.[94] To codify the new order, Chile drafted a new constitution in 1980 that not only was inspired by Hayek's legal treatise outlining a market society free from democratic constraints, but even adopted its name – the Constitution of Liberty.[95]

Back in the USSR

In our effort to understand how Half-Earth socialism could work in practice, we have undertaken a tour of socialist planning theory that has travelled from Leningrad to Vienna to Novosibirsk and finally to Santiago. At this point, we return to Siberia some two decades after Kantorovich's work at CEMI began. In the 1980s, Olga Burmatova, a cybernetician in Novosibirsk, sought to reconcile economic planning with protection of Lake Baikal, more than 1,400 kilometres to the east. By following Burmatova, we discover an experiment in planning that combined decentralized control within a total plan – something that Beer achieved with his viable system model – with explicit protection of the environment. By far the world's largest and deepest lake, Lake Baikal is home to many unique flora and fauna, including an endemic species of seal (a surprising animal to find thousands of kilometres from the nearest ocean).

Burmatova was concerned about the environmental threat posed by the Baikal-Amur Mainline, a new railway that would cut through the permafrost to link Siberia's rich natural resources to Moscow. Even worse was the Northern River Reversal Project, another mega-project that would have diverted Siberian rivers to irrigate fields in Kazakhstan – much like the disastrous 'Virgin Lands' campaign killed the Aral Sea. Burmatova realized that incorporating environmental data into economic planning was necessary to protect this unique ecosystem.

Burmatova worked within the planning framework of 'territorial production complexes', which were as close as the Soviets got to something like the viable system model. According to historian Diana Kurkovsky West, territorial production complexes were based on the insight that 'local resources, production, goods, and population had to be considered as existing simultaneously within various networks, on many scales, embedded in several economies, and therefore in a host of complex relationships to each other'.[96] Burmatova thought that this kind of model naturally allowed for the inclusion of environmental concerns. She was especially interested in the local impacts of industrial projects: 'the local level of analysis … gives one the opportunity to more closely study and consider the specificities of individual plants, natural resources, and conditions of a given territory from the point of view of nature conservation and protection'.[97] In her 1983 monograph, she laid out hundreds of pages of dense mathematics accounting for environmental problems such as soil, atmospheric, and water pollution, showing how they could be used to determine whether a project was worth its environmental cost.

Like other planning schemes inspired by control theory, Burmatova's environmental model required an enormous amount of continuously updated data, which she recognized as a challenge. As West summarizes, 'Burmatova believed that the answer to resource depletion lay in a system of continual informational feedback, performed in real time, in the future world of a fully computerized command economy.'[98] If the cybernetic system lacked this level of data throughput, environmental issues would inevitably go unrecognized and uncorrected (though human interpretation is still crucial, as the history of the ozone layer makes clear). That the Soviet Union lacked the kind of dense data networks and algorithms required for cybernetic environmental planning was a source of continual frustration for Burmatova.[99] Now, in the age of satellites and ubiquitous environmental sensors, climate and weather models have access to information Burmatova could only have dreamed of, and meteorologists and climate scientists have devised clever ways of using that data to control their models.[100]

It is no coincidence that environmental scientists eventually solved Burmatova's informational problem, and in the process developed the global environmental models that we rely on today. Indeed, the Soviet Union was a crucial player in the development of climate science. Remember how Gosplan refused to give data to Kantorovich and other reformers, to stymie their simulations? Rather than battle the intransigent bureaucracy, many Soviet scientists went ahead and built massive models anyway, but instead of simulating the USSR's economy, they modelled the planet's biosphere and climate. In 1977, the Computer Centre in Novosibirsk launched a programme to build what is now called an Earth system model (ESM), an enormous computer simulation that emulates the physics and chemistry of the entire planet, predicting everything from weather and climate to air quality and environmental health. Completed in 1982, this ESM allowed scientists to safely intervene in political debates without provoking a backlash from elites. For example, the 1980s controversy over the possibility of a postbellum 'nuclear winter' was sparked by ESMs, thus forever changing the politics of atomic weapons.[101] These enormous models also facilitated co-operation between modellers across the Cold War divide, much of which took place at Austria's International Institute for Applied Systems Analysis (IIASA). IIASA was founded in 1972 and directed by Dzhermen Gvishiani, the son-in-law of Kosygin and close ally of Nikolai Fedorenko, the first head of CEMI. Today, it houses many of the most influential IAMs in climate politics.

If the Soviets themselves worked on Earth system models to obliquely discuss the politics of nuclear war and environmental degradation, it is worth considering this scientific field – enormously active today – as the proper inheritor of the lost art of central planning. For example, Beer's structure of a central office updated by a wide network of local data sources has a long history in climate science. As historian Paul Edwards argues, climate science and meteorology were the first fields that collected and processed global data in near real time. To make even a banal statement like 'The world has warmed by 1°C since 1850', scientists need to compile data from a vast global network of ground observation sites, weather balloons, research vessels, and satellites, then feed that data into enormous physical and mathematical models to form a coherent world picture. Edwards calls climate science a 'vast machine' in that it is 'a sociotechnical system that collects data, models physical processes, tests theories, and ultimately generates a widely shared understanding of climate and climate change.'[102]

When you check the weather in the morning, you are tapping into knowledge produced by a global network of supercomputers, all constantly running models containing everything we know about the physics of clouds and water and heat. Because the weather is a chaotic system, even a model with perfect physics will quickly spin away from reality due to the famous 'butterfly effect'. As a result, weather models are constantly corrected by a global data network. Everything we know about weather and climate is a fusion of real-life observations and idealized models, much like our incomplete knowledge of nature in general.

What do ESMs have to tell us about controlling a large, complex system? As Kantorovich noted in his vision of multi-level linear programming, a planned society needs a way to balance an overarching national or global vision with sensitivity to local conditions. Luckily, atmospheric scientists have developed a whole scientific framework for doing just that. Using supercomputers, scientists run models of the entire world at coarse resolutions, overlooking some of the small-scale dynamics of nature that are infeasible to include in such a huge simulation. Local scientists then use those large models to shape their base assumptions of how a smaller spatial area works, and run their own more detailed model on that region (e.g., a continent, a nation, or a city).[103] This system, which climate scientists call 'downscaling', recalls Beer's viable system model, in which there are interconnected and hierarchical modules that govern at different levels of complexity. Gosplant would need to have similar capabilities. A global plan needs to be flexible enough to respond to catastrophes and adjust to ever-evolving social mores and science, for Half-Earth socialists plan without full knowledge of nature.

As it turns out, climate scientists and meteorologists have also figured out how to continuously update their models with new information, having developed extraordinarily powerful algorithms in a field known as data assimilation. Environmental models can be adjusted using near real time or historical data, collected by a diverse group of scientists around the world, ensuring that the simulations never spin too far away from the situation on the ground. Weather models rely mostly on satellites, balloons, planes, and networks of sensors to supply the data they need to constrain the simulation. The

Gosplant social engineers might consider more diverse data sources, such as wireless trackers on wild animals, summaries of factory output, updates on infrastructure projects, and reports and observations submitted by citizen scientists. This flexibility does more than just improve the model's performance: as planners monitor their simulations and analyse when the data and the models diverge, they can use such errors as an opportunity to update their assumptions. In a historical twist, Kantorovich's work on transportation theory – part of his corpus on linear programming – is used by meteorologists to update their models on the basis of new information. Rather than moving mounds of dirt, as Kantorovich originally imagined, scientists move around abstract probability distributions to achieve the ideal synthesis of model and data.[104]

Earlier planning theorists would have immediately recognized the worth of these observing-modelling systems. Cybersyn included a primitive algorithm that compared real time economic data with past data, checking for anomalies and allowing for a deeper understanding of the dynamics of Chile's industries.[105] Contemporary economic planning can draw on techniques from meteorology to do a far better job. Neurath would have adored the data assimilation we have today. In 1913, he bemoaned the lack of industrial statistics and feared that missing data could undermine central planning. 'It is either difficult or impossible to link or compare various statistics', he wrote. 'At countless places economic statistical studies are made. Cartels, unions, individual bodies alongside state and municipal authorities that have their own statistical offices, all gather statistical data. But generally here the right hand does not know what the left hand does.'[106] The data density of the contemporary world, paired with the algorithms climate scientists have designed to handle it, greatly expands planning capacities.

Every element of Half-Earth socialism's 'vast machine' of planetary calculation is based on already existing technologies.

The central algorithms in the model would take advantage of many of the insights and engineering designs that climate scientists have spent decades developing. Its tiered structure could draw on the nesting pattern found in environmental and atmospheric models, with global and local simulations constantly interacting and updating one another. Those algorithms would allocate simple, existing technologies like solar panels, organic agriculture, wind turbines, and public transit – and, in doing so, would help us avoid gambling the future of the planet on untested technologies like SRM that would create new problems we can ill afford. All this data does not mean that we fully know nature, only that Half-Earth socialist planners would have access to the vital signs of the planet so they could modify humanity's interchange with nature when necessary.

The Limits of Planning

Socialism is society emancipated from the relentless, unconscious, and irrational power of capital. Living in a planned society should feel *better* and *freer*, with a sense of solidarity and freedom from the threat of poverty. Democracy and meaningful work are not mere side effects of a socialist economy but central for planning to function, because no one will work hard on a project they don't believe in. If people stop believing in socialism, the system will collapse, as was made clear in the strangely sudden and peaceful denouement of 1989. If people do not have a say in how the economy is consciously managed, then even the most technical and ecological planning will degenerate into tyranny. This is why any attempt to revive the utopian dream of a consciously planned economy, such as Half-Earth socialism, must reckon with the failures of past socialist societies.

János Kornai, a Hungarian economist, had a hard-nosed but fair theory of actually existing socialism. Indeed, his work achieved the rare feat of garnering respect on both sides of the Iron Curtain.[107] In short, he argued that socialism inevitably led towards a *shortage economy*.[108] The causes were myriad but ultimately boiled down to the way individuals behave in a planned economy. While capitalist firms are motivated by maximizing profit, socialist firms were primarily motivated by satisfying output quotas set by a bureaucracy. This led to pathological behaviours, such as duplicated capacity, where firms created their own additional productive units to secure small batches of supplies because they did not trust other factories to produce necessary intermediate goods. This made output more reliable but negated any gains from economies of scale. While an inefficient capitalist enterprise is punished by going under, the inefficient socialist firm had only the Party to fear – and the Party had interests other than efficiency (as Kantorovich knew only too well). The underperforming but favoured firm would always receive help, while a well-run factory could be shut down if it antagonized a powerful patron.[109]

Kornai contrasted socialism's congenital problem of shortage with capitalism's tendency to *surplus*. Profit encourages firms to produce more than what is needed. Indeed, capitalist growth rates have slowed over the past half century in part because the most dynamic sectors, such as manufacturing, have productive capacities that far outstrip demand.[110] Even Larry Summers, a paragon of the establishment, recognizes capitalism's tendency towards 'secular stagnation'.[111] In fact, the environmental crisis itself could be considered an inevitable consequence of capitalist overproduction, where too much of nature has been humanized, to the point of destabilizing capitalism itself. From this point of view, the most natural solution may very well be a socialist system that errs on the side of shortage.

Socialism's unification of the economic and political spheres – which are separated in capitalism – is essential now to overcoming the environmental crisis, but such an act involves considerable dangers. Duplicated capacity, forced substitution of inferior production material, and an inability to correct inefficiencies will plague a planned economy if they are not swiftly addressed. Restricted information can also be fatal. Yet planners as a group can be insulated from such inefficiencies and informational problems. Gosplan, as we have seen, refused to share vital industrial statistics with reformers like Kantorovich. Moreover, its method of 'material balances' was ad hoc and unresponsive to changing conditions, but Gosplan's bureaucrats had little incentive to improve as long as their redistributive – and thus political – power was secure.

Although we think a wisely designed, democratic, and epistemically humble planning system would solve many of the problems that twentieth-century socialist regimes faced, some of the ills Kornai diagnoses would undoubtedly afflict Half-Earth socialism too. Capitalism comes with an embedded form of coercion that drives growth: the threat of unemployment, homelessness, and starvation for those unwilling to sell their labour, and bankruptcy or hostile takeovers for any firm that fails to make the going rate of profit. The socialist economy, however, will necessarily have weaker powers of coercion. If we believe that every human being deserves food, shelter, health care, education, and electricity, then we cannot force people to work by withholding those needs. Everyone will be motivated instead by positive incentives that lack the knife's edge of immiseration, such as social obligation, personal satisfaction, pride, leisure, and even modest material bonuses.

Moreover, one can concede to the neoliberals that even the best plans based on the most up-to-date information will not be as dynamic as the price system. As a result, the tendency towards shortage that Kornai diagnoses will likely emerge in some form. Socialism involves certain trade-offs that are inseparable from the system itself, just as the dynamism of capitalism goes hand in hand with inequality, unemployment, and ecological devastation. As Kornai writes, 'It is also impossible to accept with pleasure the beneficial effects of [socialism] and to escape entirely from those consequences which we regard as disadvantageous.'[112] Replace 'socialism' with 'capitalism' in Kornai's quote, and you have an argument for why revolution is necessary to avert ecological collapse. If we truly want to overcome the environmental crisis, we must transition now to a socialist society founded on ecological principles, and do our best to manage the inevitable pitfalls of the new system.

However, if Half-Earth socialists are sufficiently devoted to democracy, and are willing to make life under the new regime into something beautiful and desirable in its own right, then the problems of bureaucratic pathology and motivation will not be fatal. Every part of Half-Earth socialism we have proposed should be seen not as an unquestionable truth but as a starting point for a deeper discussion of how socialism should function in an age of ecological crisis. Our imperfect knowledge of nature and society will lead to blind spots that the Half-Earth socialist planning bureau cannot always fix. These weaknesses are to be expected in any plan that confronts the catastrophes of the Anthropocene, but our hope is that Half-Earth socialism will differ by producing a society which constantly revises itself towards a more just and environmentally stable civilization through conscious choice. This does not mean that creating a global utopia will be easy. Yet, a socialist society confronts this challenge with open eyes, rather than trusting the mythical powers of the market. In such a struggle lies the possibility of human freedom on a self-willed natural world.

4

News from 2047

> Go back again, now you have seen us, and your outward eyes
> have learned that in spite of all the infallible maxims of your
> day there is yet a time of rest in store for the world, when
> mastery has changed into fellowship.
>
> —William Morris

William Guest woke up to sunlight streaming through the
window. He stood up slowly, as if pushing a lifetime of sleep
off his shoulders. The light filtered through his tousled hair
and tangled beard, revealing a few grey hairs on the fringes.
Still groggy, he saw only that the room was small and faced
the sunrise, with tufts of green emerging from pottery by the
window. At least something is enjoying the morning, he thought,
before disappearing down the hallway to the washroom.

Seconds later, he returned and stood in the doorway, his
forehead furrowed in confusion. The floor was made of wide
pine planks, darkened and smoothed from years of use, as if
they had been recycled from an old house – but he was certain
he'd gone to sleep in his vinyl-lined apartment. The bed had a
simple frame but was covered in a now rumpled quilt stitched
with interwoven organic shapes. Leaves, stems, flowers, and
birds coursed together in patterns of increasing complexity,
radiating the warmth of a summer picnic throughout the
room, colouring the sunlight green here, red there. Following the beams of light towards the window, his eyes briefly
stopped at a small desk, covered by a stack of thickly bound
library books. As his gaze wandered across the creased spines,

he noticed that the window was made of unusually thick glass, like a ship's porthole. The walls, too, seemed strangely thick, forming an enormous windowsill that could support a flowerpot overflowing with basil and thyme on one side and a small purple cushion on the other. He sat down, folding his legs to his chest, and peered through the window. Just below him, twining stems of countless beans climbed up a trellis overlooking a small garden of eggplants, peppers, and tomatoes. He recognized the native flowers nestled among the food crops: Solomon's seal, New England aster, and butterfly weed, which the bumblebees and monarchs seemed to be enjoying. Beyond the garden, rolling hills were planted with bushy green vegetables, interlaced with towering wind turbines and equally impressive oaks. To his left, the building itself soon merged with a hill, and the garden rose along with the land to blanket the roof with a thick pelt of earth. On top, he saw a large, contorted beech tree with deep purple leaves. He could have sworn it was identical to the one behind his apartment, though it was thickened with age.

Guest wracked his brain for memories of the night before. Nothing that unusual. He had spent an evening at a bar in Amherst with six fellow socialists, who, naturally, represented six different radical traditions, including four strong but divergent anarchist opinions. While seeing his friends had lifted his spirits, Guest remembered yesterday's undercurrent of despair. It was a bad day, with the Left's champion losing a primary contest to the establishment candidate, the same day that the last white rhino died. It seemed that much needed to be done to save the biosphere, redistribute the world's wealth, and create a real future to replace the unending capitalist present. Perhaps he was still dreaming, a sort of solace offered by his anxious imagination. Yet, try as he might, he could not wake up.

Guest walked out his door and down the hallway, which seemed to him like an unusually whimsical college dormitory. Many other doors lined the hall, identical to his own, save that most had been intricately decorated by their residents. One was covered in a glass mosaic that must have been cut from old jars and bottles, given the colourful curved fragments freckled with raised text. There were a few hallway bathrooms and a kitchen, as well as several living rooms ranging from nooks fit for a solitary reader to expansive chambers with plenty of broad armchairs. Some had wooden toys and picture books strewn about, while another had a piano, a couple of guitars, and a kora. He peeked into the kitchen. It was as well equipped as a restaurant, but the wooden countertops – thick, oiled oak – took away the sterile feeling that he remembered from his old job at a fast-food joint. Guest walked inside and stopped by an enormous steel pot splattered with tomato sauce, sitting on top of a smooth, black induction cooktop. An older woman wandered into the kitchen, her hair a short mop of tight, greying curls, and jumped back a little when she saw him standing over the dirty dishes.

'I didn't realize anyone would be up this early', she said sheepishly. 'I meant to wash up last night, but we played cribbage until late.'

Guest just stood and looked at her, unable to think of anything to say in response. Eventually, he mustered a weak smile and a nod.

'Don't tell He-Yin, but we took a little bit of dandelion wine from the barrel in the basement pantry', she said with a wink. 'It may not be ready yet, but it was good enough to put me to sleep faster than I expected.' She started to wash the dishes, filling the sink with water and beginning with the cups and plates, which had been scraped mostly clean into a compost bin.

'Well, if you want to help me out, I'll do the same on your day', she said to Guest, gesturing towards a brightly coloured rag that looked as if it could have come off his quilt. He dried as she washed dishes from cleanest to dirtiest, ending with the big pot.

'What's your name? Are you new here? I don't think I've seen you around', she said, her arms elbow-deep in bubbles. 'I'm Amara, by the way.'

'William', he said. 'And I guess I am new here, though I don't really know what "here" is.'

'I think I know what you mean', Amara laughed. 'I'd never seen anything like this, either, when I first came out here. I only meant to stay for a couple of months. A break from Boston, plus I heard they needed help with the harvest. But I really took to it. There's something wonderful about working with your hands.'

She paused and looked out the window. 'I've been living in the dorms for two years now', she said. 'Loved every second of it, but I'm not sure how long I'll stay.'

'Why not?' Guest asked, a note of concern in his voice.

Amara raised an eyebrow. 'Did you not see the last plan? The summary is in the library downstairs.'

'The plan?' Guest muttered in confusion, but Amara didn't seem to notice.

'They're letting most of Massachusetts rewild, since the soil's not as good as out west. Right now some vegetables still make sense to grow at scale in New England, since it saves a bit on transport, at least until they finish the electric rail lines out in the Midwest. Once that project is done, though, this farm's days are numbered.' She paused for a second. 'I think I'll apply to work in Iowa when the time comes, unless these hands have gotten too old. But I'll miss all of He-Yin's harebrained fermentation experiments.'

They silently put away the clean dishes, then scrubbed the wooden counters and black stovetop of all specks of tomato sauce. It smelled like summer in the room. Fresh herbs overflowed from pots by the window, and heirloom gourds with flamboyant stripes were piled up in bowls by the refrigerator. Guest spent the whole time wracking his brain. How could he ask Amara basic questions about this place without making her think he was crazy? He certainly felt crazy, and hiding that would be a delicate task. Quickly, he settled on a story.

'You know, I was born in Amherst, but I've been abroad for a long time', he said, watching her face carefully so he could make any necessary adjustments to his story. 'I came home not too long ago and made my way here yesterday. I don't think anyone but you knows I'm here.'

Amara had been polishing the dining table but paused on hearing this. Guest held his breath until she burst out laughing a few seconds later. 'You must have been a long way away!' she said. 'Well, if you'd like, we can get you registered here. Always need more hands. And Edith will be able to answer all your questions. She's the farm's rep to the regional council.' Guest sighed with relief as he followed Amara down a staircase just outside the kitchen.

The walls of the staircase were painted with elaborate murals of fields, forests, and rivers. The landscapes were in several styles, done with varying skill. 'The dormitory spent a weekend painting these last winter', said Amara. 'It might have looked better if we'd invited folks from the artist's colony up north, but we voted to do it ourselves, have a bit of fun. I did that lake over there, with the moose, but I never managed to get the antlers right.' Guest looked over where she was pointing and saw a blob with sticks coming out of its head. 'There are some real talented people here, though. You should chat with Morris. He's the one who made the quilts.'

'Is he a full-time artist?' asked Guest.

'Nobody is a full-time anything around here!' Amara said with a laugh. 'You know how Marx said that one day we'll all "hunt in the morning, fish in the afternoon, rear cattle in the evening, criticize after dinner"?[1] Well, we're vegan so we don't do that stuff literally, but you know what I mean.'

Guest gave her a startled look, but she seemed not to notice.

'Our main job is to tend to the vegetables', she continued, 'but that's not too much work. Harvest and planting season are busy, though we grow so many different species that the work is spread more evenly throughout the year than in monocrop farms. There's always maintenance to do, but most of the time we only farm a few hours a day. That leaves plenty of time for second jobs and hobbies. He-Yin cooks and makes moonshine, but she's also a biologist focusing on apex predator reintroduction. Morris weaves and fixes up old clothes and shoes, while Edith is always in the library poring over the regional and global plans. Plus there are study groups on everything from architecture to mathematics. It's a very vibrant place.'

'And what do you do?' asked Guest.

'Oh, I dabble in lots of things', she said. 'I like to keep up with Edith so I can send my comments on to the regional planners, and lately I've been helping He-Yin crunch some numbers about the deer population. My parents spoke Igbo, but not with me so I never picked it up. I've been practising with some climate refugees from Nigeria at the university in Amherst.' She paused. 'Don't tell anyone, but I'm also trying to write a mystery novel, since painting does not seem to be my forte.'

The downstairs hallway looked much like the rest of the building, except the shared rooms off to the side were much bigger. There was a library stocked with shelves of books and velvet armchairs, but also some long tables scattered with a few worn but sturdy laptops and phones. The bedrooms seemed bigger than upstairs, and some of the doorways were decorated with children's crayon drawings. Amara stopped in front of a door marked DR. EDITH LEETE, along with the words REPRESENTATIVE FOR FARMING DISTRICT 11. She knocked.

The door swung open, revealing a spartan room decorated only with two computer monitors, piles of textbooks on everything from soil ecology to educational theory, and an enormous binder made even thicker by all the dog-eared pages and colourful bookmarks poking out. Standing in the entryway was a woman whose straight red hair was pulled back in a tight ponytail, and who held a mug of black coffee.

'Good morning, Amara', she said curtly, and turned towards Guest. 'Are you visiting someone in the dormitory?'

'Actually, he's new here', said Amara, before Guest had the chance. 'This is William. He kind of wandered in. Not registered in the system, and he has the softest hands I've seen in a while, but a good worker.' She winked at Guest. 'He helped me out this morning in the kitchen. I was thinking we could get him assigned to our unit.'

Edith frowned. 'We haven't done that sort of thing in a while. The whole point is to centralize these kinds of decisions. Make sure jobs and workers are optimally matched and all.' She paused, then sighed. 'But, even ten years in, I suppose there are still plenty of kinks.' She hung on to that last word as she stared at Guest's messy beard. She waved her hand at a corner chair and said goodbye to Amara.

Guest sat down as Edith pulled up a database on the computer. 'Any agricultural experience?' she asked.

'Nope', replied Guest. 'I'm a programmer, but in undergrad I took a class in botany once, if that helps.'

'It doesn't', said Edith flatly. 'We'll assign you to the school. Your first month will be intensive training, which will get you ready to take on most of the field chores. We always encourage people to take more classes afterwards, especially if they want to be a shift leader or contribute to the local or regional plans. Your computer skills would come in handy there.'

'I was telling Amara this, but I don't quite understand what you mean by the plans.' He paused, choosing his words carefully. 'I've been away for a long time.'

She eyed him sceptically, but the confusion on his face must have seemed genuine. 'Well, the system is new, and large parts of the world are still transitioning from the old regime, but the basic pieces are in place. Do you at least know about the Gosplant office down in Havana?'

Guest shook his head. Edith raised her eyebrows.

'Well, the parts of the world that have joined the Half-Earth bloc all provide data, technical expertise, and proposals to the central planning bureau in Cuba', she said, leaning forward in her chair. 'They have had the most experience with planning, decarbonization, and organic agriculture, so I guess it made sense to put the bureau there. You should really see a picture of that place. It's enormous.' She spread her arms as far as they could go, then turned to her computer and brought up a picture of a renovated Art Deco building with trees growing on its wide, flat roof.

'What do they do there?' said Guest, amused by how excited Edith was getting about an office building.

'Thousands of people work there from all over the world', said Edith, her sparkling eyes betraying her otherwise tight-lipped demeanour. 'They use these massive supercomputers to make a series of global plans simulating snapshots of the future: say, five detailed blueprints of the planet five years out, ten for the coming decade, and a couple dozen for the next quarter century.'

Guest looked at her quizzically. 'How do they do it? How would they have enough data for the whole world, let alone several possible futures? And even if there were enough, what kind of computer could process it?' he blurted out, and then immediately worried about coming off as hostile.

'You're right that the data requirements are staggering', she said, and Guest was relieved to see the smile lines around her eyes deepen. 'The thing is, it's not necessary for all knowledge to be in one person's head or one office. The planners in Havana have a lot of information at their fingertips, but they only need a rough picture of the globe, like maps of biodiversity and climate, estimates of worldwide food and energy needs, and constraints on resource use. Those were available long before the revolutions began.'

'Revolutions?' Guest interjected.

'New societies are not born on their own', Edith replied with a wry grin.

'But—' Guest managed to suppress his bewilderment, and he tried to stay on point. 'Surely that's not good enough! It's nice that the planning bureau is thinking about big issues, but how does that help you decide how many black-eyed peas to plant, or where to build the next train line?'

'That's why the global model is coarse!' said Edith as she opened the binder, her voice raised in excitement. 'Here, take a look at the maps from the latest short-term blueprints.' She flipped to a particularly dog-eared section and stuck her finger on a picture of North America, divided into a grid covered in multicoloured squares. It looked a bit like a pixelated version of the storm maps on the Weather Channel.

'This is the plot for Pasture Reclamation Plan 5-F', she said, in a now enthusiastic tone that reminded Guest of his brother when he talked about his favourite baseball players. 'The folks down in Havana made a few different plans for the next year assuming different energy quotas, ensuring in each plan that nature is always respected and no one is left behind. Plan 5-F is the most austere, keeping consumption at about 750 watts per person, which includes all of the industrial production and social services required to keep society running. Once they send out the global blueprints for many possible futures, regional and local planning offices make their own plans that both meet the conditions of each of the global plans and respond to local concerns.' Edith gestured at the piles of textbooks, scientific papers, and binders covered in red ink.

'I represent this farming complex of about a hundred dormitories, five thousand people or so, and my main job is to talk with the people here and use their experiences and needs to help form some local blueprints to propose', she said proudly. 'For us,' she continued, 'Plan 5-F means we can let a lot of pasture return to wilderness, rather than planting more of those terrible biofuel plantations. There's a lot of support for that in New England, what with the big hurricane of 2029 and the ghastly avian flu pandemic.'

Edith seemed to withdraw into herself after mentioning the outbreak. Guest squinted at the binder, unsure what to make of it. It was unbelievably detailed, like an encyclopaedia-sized scientific paper.

'We're more in favour of energy austerity here than in many places', she said after a long pause. 'That's probably because these dorms are passive buildings, needing little energy to cool or heat, even in the dead of winter. We can get away with those terrible electric heaters on the worst nights, because almost no warmth gets lost.'

'Really?' said Guest, eyebrows raised in surprise.

'Transportation is the only thing that makes people hesitate', said Edith, a hint of frustration mixed with the excitement in her voice. 'With a quota that low, we will need serious rationing of long-distance transit, but that's a necessary sacrifice until we can bring more electrified public transportation online. I'm awfully jealous of places like Japan and Switzerland, where they already have extensive trains in rural areas. In the short term, though, we need to rely on biofuels for large parts of transportation, and those just kill the short-term plans with higher energy use. We end up planting so many energy crops that we take up too much of nature's domain. In the long term, I think we will be able to move around almost as freely as in the before times, but there will never again be something like private car ownership. It just can't work with the global plans. But from what the old-timers like He-Yin tell me, driving wasn't much fun, anyways.'

Edith started flipping through the binder again, but Guest interrupted before she found the transit section. 'What happens after you make all these plans?' he asked.

Edith put down the papers and motioned for him to follow her. They walked back into the hallway, which was now humming with morning activity. Three giggling toddlers ran by them and into the library, and Guest had to leap out of their way. 'Sorry about that', said a white-haired man with a laugh, as he tracked the kids' trajectory.

Edith shook her head and smiled, then continued down the hallway. 'Well, after having a few farm-wide meetings, I have a good idea of how our constituents will want me to vote and negotiate at the regional parliament. We also have a representative at the World Parliament in La Paz, who will take what we decide here and push for it on the planetary stage. The whole thing can still be a bit contentious, but we've always been able to come to an agreement. It's been a lot easier to reach a consensus since we set global living standards a few years ago.'

'La Paz?' Guest said, barely able to contain his surprise.

'Yes, it feels a bit provincial here sometimes', Edith answered absent-mindedly.

They walked out the front door of the dormitory into the garden. Around the side of the building, Guest saw a canopy covering a row of bicycles haphazardly arranged on the compacted dirt. A small silver pickup truck sat parked behind them, with a spare interior consisting of little more than a few seats and a steering wheel. The makeshift garage was framed on the right by the dormitory walls, and from behind by the rising hill, which was covered in wildflowers. An equally beautiful door was set into the rising meadow, decorated with intricate carvings. It opened, and Guest saw Amara come out with a crusty slice of bread.

'Hey, you two', Amara said as she took a final bite and licked her fingers. She jumped on a bike and clipped on a helmet that was attached to the handlebars.

'Hi, Amara!' said Guest. 'Where are you off to?'

'I'm heading into town', she said, smiling. 'Seeing an old friend before my shift this afternoon.'

They waved goodbye as she leisurely pedalled away. Guest's eyes drifted back towards the garden in front of the dormitory. At the centre, there was a statue of a balding man with thick glasses, with the base of the pedestal taken over by some climbing beans.

'Who's this?' asked Guest.

'Old Lyonechka!' Edith blurted, as if cheering on a mascot. 'I mean, Leonid Kantorovich. We named the dorm after him. He's a big inspiration behind this "parliament of blueprints", as we say. He described the problem of socialist governance long before our movement came to power.'

Guest read aloud from the plaque: 'The problem is to construct a system of information, accounting, economic indices, and stimuli which permit local decision-making organs to evaluate the advantage of their decisions from the point of view of the whole economy.'[2]

'Not the most eloquent guy', Edith said, 'but smart.'

They paused in the garden, enjoying the sunshine and the flowers. It was humming with life – Guest had never seen so many bees, insects, and birds. A forest extended back behind the hill, and he could have sworn he spotted the antlers of a moose peeking through the trees. Surely there aren't moose in Massachusetts, he thought.

'My hunch is that the global energy quota will be settled at more like 1,400 watts for the next five years, with the goal of boosting it up to 2,000 watts in the long term', said Edith after a while, emerging from her own thoughts. 'If that's the case, I don't think it will affect us too much. We'll just use less than we're allocated, which will free up other areas to use more power as they develop infrastructure. I just hope we don't have to grow too many biofuels. Ever since that big BECCS plant down south caused all those water and soil issues in the 2030s, I've been sceptical.'

'How would that work?' asked Guest. 'How will, say, your counterpart in Mumbai know that her region is able to use more energy to build housing or transport, just because a farm in the Connecticut River Valley kept its energy use low?'

'That's a good question', said Edith. 'So far all I've been

showing you are plan *blueprints*, which we create using tools of mathematical optimization, not unlike what Kantorovich was doing a century ago. Once La Paz and the regional parliaments work out which one to adopt, then planning offices at all levels begin booting up their Half-Earth system models.'

'Half-Earth system models?' asked Guest.

'That's a harder one to explain', said Edith, thinking for a moment. 'Let's go to the school. There's a nice display there. We'll bike.'

'What about that truck?' Guest asked nervously. He hadn't cycled in years.

'We try to save that for people who need help getting around, or for moving heavy things', said Edith, before seeing a concerned look on Guest's face. 'Don't worry, we can go at Amara's pace. It's always good to enjoy mornings like these. But I suppose we could catch the bus.'

Guest exhaled in relief. 'No, biking will be fine', he said. After all, the sun was shining, and the fields were even more beautiful than they appeared from his room. They climbed on a pair of bikes and pedalled down the road. He had always loved the sight of rolling hills planted with vegetables, but the way the intermixed crop species seemed to emerge from the woods made the place feel more like a park than a farm.

'Yes, we're pretty lucky here', said Edith, noticing his blissful smile. 'Out in the Midwest there's still lots of industrial farming. The corn that once went to cattle now has to go to biodiesel. Many other grains are still farmed on that scale too. Eventually, they want to rewild half of the Great Plains and get a few million bison back in it, but that plan always gets delayed.' She sighed. 'Even after the revolution, change comes slowly. There are some perks to the Midwest as it is, though. We can store up plenty of surplus crops and send them to places recovering from geoengineering and climate change, where yields are a lot lower. Of course, we take in climate refugees too.'

As they passed more dormitories, Guest's mind began to wander. 'What do you do in other places?' he asked. Edith looked at him quizzically. 'I mean, how do people live in places where there were already lots of buildings, like cities? You can't rebuild everything from scratch.'

'Oh no, that would be a huge waste', said Edith. 'We filled in some of the suburbs to make them denser and more communal. Besides, the abundant lawn space allowed sports facilities and small-scale farming. Other suburbs needed to go completely, like where we are now. Maintaining them required too much energy and fuel, but at least we were able to recycle the wood in these dormitories. Before, I heard they would just demolish a house and send everything to the landfill.' She scrunched up her face in disgust. 'In general, people have been very creative about reusing old buildings. One of the most common jobs assigned these days is a retrofitter. We need huge numbers of carpenters and other artisans to stuff extra insulation in the walls, split up McMansions into multifamily apartments, and generally prepare cities and towns for the low-carbon future.'

After about twenty minutes they arrived at the school. 'Shawmut College!' Edith said, proudly.

At first, the building seemed to have a design similar to the dormitories', except it was much bigger and shaded by a grove of towering oak trees. Gardens in raised beds surrounded the building, with some chairs and outdoor chalkboards interspersed between them. They stepped inside, and Guest was taken aback by the apparent chaos. There was an enormous main chamber with a glass ceiling, and tables and desks arranged haphazardly in what appeared to be makeshift classrooms. Some had dirt and leaves scattered over the surface, others blocks, still others electronic components. Brightly decorated rooms of every shape and size branched off the main hall. Edith walked directly into a room on the right, and Guest hurriedly turned to follow. Inside, there were about a dozen desks, all angled towards a big screen. Edith pressed a few buttons and the screen booted up, revealing a map of the globe surrounded by several charts and graphs.

'Once we figure out the plans we'd like to follow', she began, 'the central planners collaborate with mathematicians and worker-representatives to devise a dense group of interlinked differential equations modelling how industry, construction, and agriculture will need to change over time to meet the goals of the global plan.' The screen showed various animations of line graphs, all moving in different directions. Guest noticed that whenever an orange line ticked up, a green line seemed to tick down, and he was about to ask Edith about it. She spoke before he could. 'You said that you studied programming, so I'm guessing your math skills aren't bad, no?'

Guest nodded. 'They're not too shabby, but take it slow just in case. I work mostly in social media.'

'Good', said Edith with a laugh. 'We've been a bit surprised by how well our students have been doing in math here, especially the older ones who take classes in their spare time. Never thought they were any good at it, but I think they just needed to see the big picture. Socialism will fail if we cannot plan effectively, and there is no way to plan without math. Politics are a good reason to practise your integrals.'

She turned back to the screen. 'The Half-Earth system model will need to simulate the resource flows required to build thousands of kilometres of track and construct millions of batteries, calculating throughout the impact on environmental systems. Each change in one part of society affects all other parts, and the environment too. In more rural areas of New England, for example, our energy quota is constrained by a lack of train lines, which forces us to use less-efficient biofuel buses to make up the gaps. As new rail stations open, an interlinked equation determining the energy quota can begin to slowly increase allowed use.' She pointed at a line marked with yellow thunderbolts that was slowly rising, as a purple line with a cartoon train rose even faster.

Guest must have looked lost, because Edith started laughing. 'Sorry, I get too into this stuff', she said, rubbing her arm with her other hand.

'No, it's really cool!' said Guest. 'I'm just still not sure how it works in practice. How does all this data get collected and combined with the original plan?'

Edith thought for a second, then smiled. 'Say, for example, that a public transit project is falling behind schedule in some regions', she said. 'That's not really an example. We were slow to build these things before the revolution, and we are slow to build them now.' She paused, as if expecting a laugh. Guest forced a chuckle, but the pain of riding slow, bumpy Amtrak trains was real for him in a way it wasn't for Edith.

'Anyway,' she went on, 'planning offices could use discussions with workers and reported data to understand why the project fell behind. Perhaps the problem is a bottleneck in steel-recycling plants, and more plants are needed than the original model expected. This is no problem for the Half-Earth system model. We use methods borrowed from climate science to adjust the model back towards reality.'

'But what's this new plan binder you've been talking about?' asked Guest. 'I thought we already were in the Half-Earth system model stage?'

'We're always making new blueprints for the future', Edith replied. 'That's the nature of democracy: permanent dissatisfaction with the present. We're constantly learning, and there are always new concerns. For example, it seems likely that fishing is going to be more or less banned in the next parliamentary session, partly because we know more about the relationship between marine animals and the carbon cycle than before, but also because the animal-rights movement has grown a lot stronger. Anyway, once we've finished the regional blueprints for Massachusetts, we'll open them up to the public and have a good discussion, I'm sure.' She winced a bit, and Guest imagined that a 'good discussion' might involve a bit of yelling.

They powered down the monitor and stepped into the main chamber, standing in silence for a while. Guest's eyes wandered over to a set of questions on a chalkboard beside a table covered in soil samples. He frowned as he tried to read from a distance:

UNIT EXAMINATION (Grade five)
'How well do you know your home?'
- Define the limits of your bioregion. Be able to justify the boundaries you choose.
- How many days until the moon is full (plus or minus a couple of days)?
- Describe the soil around your home.
- From what direction do winter storms generally come in your region?
- Name five trees in your area. Which of them are native?
- From where you are reading this, point north.
- Which spring wildflower is consistently among the first to bloom where you live?
- Were the stars out last night?
- Name seven prominent landforms in your region. Whose language is used for those names?
- Give five aspects of your life that are independent of your bioregion. Where are they supported by Earth elsewhere?

Edith saw him staring and laughed. 'Pretty tough test, huh?'
'Yeah', said Guest. 'I've never seen anything like it.'
'We got that test from Axe Handle Academy, a curriculum proposed way back in the 1980s as part of the Alaska Native Knowledge Network.[3] It seemed right for what we're trying to do with Half-Earth socialism', said Edith.
'I don't think I could answer more than a couple of these questions, and I'm not ten', said Guest.
'We still haven't fully decided what socialist education should be', Edith said thoughtfully, 'but we have some principles. It

should be free, lifelong, and critical. There are lots of specific things people need to know, since our society relies on everyone's participation. Mathematics and natural science are indispensable for planning, but they're far from the only knowledge required. There needs to be a sense of appreciation for the fragility of nature, and a deep respect for the cultural wisdom of the past and present.'

They paused, looking back at the exam questions.

'Reforms in the old days could only take us so far', Edith said. 'Calls for people to live more ecologically or for society to become more democratic never could take hold if they conflicted with the concrete realities of how people lived and the interests of capital. We've still got a lot to learn about what human nature is, even in this new world.'

Guest sat down at one of the tables, brushing some electronic parts aside so he could rest his elbows. 'Amara said this place was going to be closed down soon', said Guest. 'I admit it makes me sad. You have such a paradise here.'

'We are building a new society, one that can exist safely on this planet for centuries while lifting up everyone', said Edith. 'That does not happen overnight, and it demands some temporary measures. This farm is one of them, and always was meant that way.'

'But what will happen to the dormitories, the fields, the people?' asked Guest.

'That's up to us, really', said Edith. 'In the long term, the global planners want much of this place rewilded, although some tree plantations are required in most plans to help meet lumber demand. They don't really care what we do with the dorms. Coarse plan, remember?'

'So what will you do?' asked Guest.

'We've started talking about this around the dinner table', said Edith. 'The dorms are built from standardized parts, so we could easily disassemble them and use them in other buildings. But I personally like Amara's idea, which is to turn this place into a combined wilderness management site and educational centre. She's in talks with universities and Indigenous leaders in the neighbouring district to turn this whole place into something like a giant Axe Handle Academy. Amara's a polyglot, really good with languages, though she'd never brag. She is just thrilled at the thought of the Indigenous language revival movement having an institutional home here. There's a radical promise there of a different kind of conservation, led by the local Nipmuc nation, with culture and science fused so that they build each other up. That movement has already taken root farther north. Did you hear how Canada elected its first métisse general secretary, Louise Riel?'

Guest shook his head and smiled. They wandered out of the hall and into the sunshine. Guest blinked as his eyes adjusted to the light of high noon. Edith spotted someone on a bicycle zipping by. 'Hey, Carmen!' she shouted. The figure screeched to a halt, the bike falling sidewise as the rider leaped from the pedals and landed gracefully on the dirt road, a cloud of dust rising around their ankles. They bent over and grabbed the bike, carrying it over to Guest and Edith with a casualness that suggested the sudden stop had been planned from the beginning.

'Hey, Edith', said Carmen. They sported close-cut black hair framing a tanned, ageless face with wrinkles around the eyes and forehead that could have suggested the imprint of age, wisdom, or laughter.

'Carmen, this is William', said Edith. 'He's new here. I've just been giving him the tour and the rundown. Didn't even know about the bureau in Havana.'

Carmen's eyebrows rose. 'Wow, really new. What did Edith sign you up for? Probably hard labour. You know you can tell her no', they said, elbowing Guest playfully.

'Oh, stop', said Edith, rolling her eyes. 'He'll start out with general farm chores. Luckily he met Amara first, and not your lot.'

Carmen laughed and turned to Guest. 'Did Edith tell you that if you work on the assembly line, you get extra perks? Farm work is on the middle tier of the compensation charts, above the pencil pushers but below more unpleasant manual work or jobs that require high levels of training. I'll happily take on a few shifts at the factory or grab a jackhammer to rip up the old interstates if it means I get more time on the Cape.'

Edith laughed. 'Carmen thinks like such a capitalist. You should be doing those things to build a better world, not for extra time at the beach', she said in a mock scolding tone. 'But we would be nothing without such workers, regardless of their intentions.'

Carmen beamed with pride and took an exaggerated bow. 'Finally, Edith says something nice!'

Edith brushed their comment aside. 'Could you take William here with you for lunch?' she asked. 'Maybe show him the factory or something. Carmen is right that you are certainly free to take on shifts if you'd like. If it seems really appealing to you, I can change your status in the system so that you are primarily a factory worker and not a farm labourer. We're short on both, so it's no skin off my nose which one you choose.' She looked at Guest, whose eyes had wandered back to the rolling fields. 'But I think I was right to assign you to the training programme.'

Guest blinked out of his reverie and nodded as if he had heard what Edith had said.

'Okay, come with me, William', said Carmen. 'Let's see if I can pry you from the hands of those hippies.'

'Christ', said Edith, as she hopped on her bike and pedalled down the road. 'I'll see you at dinner, William!' she yelled over her shoulder.

Guest mounted his bike and started pedalling slowly, face towards the fields. He looked up and saw Carmen far ahead, a trail of dust spreading from behind the rear tyre. Guest stood up to push his pedals down and raced down the dirt road after Carmen. His thighs were burning, and his forehead grew slick with sweat. It had been years since he had had this kind of workout. He looked to the fields again and saw a crowd of people building fences, weeding, and harvesting some of the beans that had ripened early. He marvelled at how healthy and relaxed everyone looked.

'Hey, William,' Carmen yelled, 'stop day-dreaming, we're at the factory.' Guest turned and saw they were in front of an enormous red brick building, covered in solar panels, with more panels spread in an arc nearby.

'Welcome to the solar panel plant!' said Carmen, leaping off their bike without bothering to brake, landing with the same nonchalance as before. 'Doesn't make a ton of sense to have it here, honestly. They don't mine much silicon in New England, so we have an inefficiency from transporting all the material up here. But we do have lots of industrial capacity and technical know-how, so the planners calculated that some production here would be optimal.' Carmen was already walking through the solar panels and towards the door of the factory, and Guest once again had to hurry to keep up.

'Are we working a shift?' asked Guest, struggling to speak as he caught his breath. 'I don't know the first thing about how to make a solar panel.'

'Nope', said Carmen. 'Just getting some lunch in the mess hall. I picked up an afternoon shift, but I'm sure someone will be heading back towards the farm afterwards that you can follow.'

They walked in the door and into a white hallway, lit by bright LEDs. A group of people wearing blue gowns and surgical masks passed by, making the place feel a bit like a hospital. Through some windows, Guest could see the manufacturing floor, complete with vats of chemicals. 'Not quite as romantic as the fields, is it?' said Carmen. 'If we don't keep this place super clean, then the panels will be less efficient or might fail entirely.'

Guest frowned. 'I always thought socialists would have robots doing the unpleasant work', he said.

'Socialism isn't magic!' said Carmen, snorting involuntarily, then covering their mouth in embarrassment. 'If anything, taking out the profit motive slowed down automation. Difficult jobs will be with us for a very long time.'

'Why does anyone work here, then? This is socialism, right? You're not going to starve or be homeless if you just stay at home', said Guest.

'Now you're the one thinking like a capitalist!' said Carmen. 'All caught up in imagining the individual self-interestedly responding to incentives. Of course, there is a bit of that going on. The planners and the councils all understand that some jobs are harder or less appealing than others, and that getting enough workers to do them will require more than prestige or a sense of duty. Edith mentioned the extra vacation time, and I do like taking the train to the beach and staying in one of those fancy houses they converted into a resort. Plus the extra credits and priority housing don't hurt.'

Carmen paused and smiled, thinking about the sand and the salt breeze. 'But the main thing for me is that the work time is so low. I'm only expected to work here four days a week! The farm workers usually have chores five days a week, though it fluctuates with the season. But I get bored and take on more shifts anyway, or help out on the farm sometimes.'

Carmen paused and smiled. 'The good thing is that other hard jobs, like mining and manufacturing, are also at the top of the charts. Way above us here. This job is pretty cushy compared to refining silicon. Before, that type of work could be almost like slavery. Now at least the people are in control, talking with the planners and the councils to figure out their work conditions and how they should be rewarded.'

They wandered into the mess hall and sat at a long table next to a couple of workers.

'This is William', said Carmen. 'He's been abroad for

a while. I swear, talking to him is like stepping back to the twenties.'

'Thomas', said one in between bites, stretching out his right hand while his left gripped a falafel pitta.

'Octavia', said the other, shaking Guest's hand afterwards. She shot a glare at Thomas, who in response started theatrically chewing with his mouth open.

'Where are you staying?' Thomas asked, after chugging a glass of water.

'One of the dormitories', Guest said, and searched his brain for a moment. 'On Farming District 11.'

They all laughed, and Guest gave them a puzzled look. 'No one talks like that, William!' said Carmen. 'Here's some free advice: just say you're working on the farm. Leave all the divisions and numbers and units to Edith.'

'I can't stand those dorms', said Octavia between spoonfuls of stew. 'Feels like there's just no privacy. The apartments are so much nicer. Our own kitchen and bathroom, without all the prying from the neighbours.'

'I still think it's crazy that you moved out when you had your baby', said Thomas, waving his falafel in emphasis. 'Parents love living there. There's always someone to watch your kid! And you don't have to cook and clean after all your meals. Like a giant extended family, though plenty of those live in the dorms too.'

'A way of making your own, chosen family, without all the old scripts', said Carmen, serious for a moment.

'I can see myself living there when I'm old', said Octavia. 'There's so many elderly people living in the dorms. They love it; you can just see it in their eyes. There are plenty of ways for them to help out the community, like watching the kids or tending the garden. So much better than in the old days, when they were shuffled away in isolation because they weren't profitable.' She paused, a look of disgust appearing briefly on her face. 'I've heard that some of the regional plans have proposed moving nurses and medical equipment into some of the dorms. Renovate some of the rooms for elder care. I think it's a nice idea. Good to know that we'll be taken care of when we get old.'

'But not now?' said Thomas.

'People are always on your case when you have a baby', said Octavia. 'So many opinions. The schools all have childcare anyway. And I want to be able to take all the time I want in the bathroom.'

'To each their own, I guess', said Carmen. 'Takes up a lot more power though', they said, with a wink towards Guest. Always making trouble, he thought.

'Not that much more!' said Octavia. 'By the way, I can't believe that those self-righteous people you hang out with on that farm – no offence, William – are pushing for a 750-watt quota. I know there's a crisis, but that's just absurd. We'll all be living in dorms soon if they have their way.'

'Octavia is practically a liberal', said Carmen, winking again.

'I am not! Stop it, Carmen!' she said. 'I'll just be voting for a representative who will push for a little more energy; 1,750 watts is not luxurious, and you know it.'

'Pretty tough to make that happen while we are still building so much new infrastructure', said Thomas. 'Tell your man he needs to step up that hydrogen research. Pull some all-nighters so we can all relax a bit more.'

'I would love to take a hydrogen-powered plane somewhere', said Carmen. 'Maybe Tahiti.'

'The beach is the same everywhere', said Octavia. 'I'm so surprised you always use your travel allotments to go to the Cape. The mountains are so much more beautiful.'

Carmen stood up. 'That's it!' they said in mock rage. They motioned to Guest. 'Seriously, let's go get our food.' They walked over to the serving line.

'No steak again, huh?' grumbled a man with a greying, curly beard to no one in particular. He stood with his tray ahead of both Guest and Carmen and looked with distaste at the entrées on offer that day. Guest couldn't understand what he was complaining about. Under the label 'country stew' was a delicious-looking pot of blue potatoes, purple onion, fragrant yellow and red tomatoes – probably heirloom – and homemade seitan sausage. But then again, since he hadn't eaten all day, he probably would have devoured anything in front of him.

'Oh, knock it off, Conner', said Carmen. 'There's a crisis on, and you know it.'

'Fodder for pigs', glowered Conner.

Carmen shook their head. 'He's fine', they whispered to Guest. 'Just a little grumpy today.'

Carmen headed back to the table, but Guest hesitated for a moment. 'Where do we pay?'

'You mean with credits?' said Carmen. 'You don't have to use those here. Factory leadership puts in all the food requests.'

Guest frowned. 'What do credits do?' he asked. 'Are they like money?'

Carmen looked at him thoughtfully. 'A bit', they said as the two of them walked to the table. 'There are food credits, which can be used at the grocery store or in cafés. Those of us in the dorms only get a few because our meals are taken care of, but people like Octavia get more since they need to shop for food. You'll have to ask her about it.'

'What about other stuff?' said Guest.

'There's another credit for all the random little things people want, but you'd be surprised how few of those you need. Since housing, food, education, and health care are all covered, there's not much left to spend on. Plus, resource-intensive items are all handled by a separate system. For example, you have to put in a request for long-distance transport. In the dorms, especially, we hardly ever use credits of any kind. Dorm leadership handles orders for the building, and you can just ask for a new bike tyre or another shirt if you need it', said Carmen as they sat down with Thomas and Octavia. 'We're talking about credits and the request system', they said, to bring the two in on it.

'Well, most people don't live in the dorms', said Octavia. 'The majority are in traditional apartments, like me, and we have to deal with the system a lot more. My oven totally died last year. So dead that even those hippie repair guys didn't think it was worth fixing. Just recycle it, they said. I did not want to hear that, since they're always short on appliances, especially the big ones, and the family units are lower priority than dorms and cafeteria kitchens. Requests take an awfully long time.'

She paused to pick up her last piece of bread and mop up the last of the stew. 'A huge pain. I know so much about oven production now, since I spent a whole afternoon on that section of the plan. Had to cook on the stovetop for a month and a half. My neighbours let us use their oven for baking cakes a few times, which was kind of them. Anyway, the system uses a matchmaking algorithm to process requests. There's an estimate of the number of ovens needed built into the giant planning model. A bump up or down in the number of requests in a month will adjust the flow of raw materials from the base model, like nichrome for the heating elements. A low-oven month might allow more production of electric heaters, for example. The finished ovens are distributed according to a priority queue. We were pretty low at first, but after a nice long wait we managed to make it to the top.'

'Another reason to live in the dormitory', said Thomas with a laugh. 'One of our ovens went out and we had a new one in a week.'

'What about smaller things, like food for your kitchen?' asked Guest, looking at Octavia. 'Or a desk lamp, or clothes, or a new curtain?' He was surprised how hard it was to think of stuff to buy, with all the most expensive parts of life satisfied.

'Carmen probably told you about food credits', she said. 'There's a bit of a market socialist system when it comes to all the little things people need. Planners experiment with prices until there's a match between supply, demand, and environmental costs. Coffee and cocoa are a tad pricey, what with transportation, but since they keep for so long they're not as expensive as tropical fruit. We get oranges from Florida in the winter, but not many. On the other hand, we've got plenty of local berries here that I never knew about before. Grains and local fruits and vegetables are the cheapest, and in the winter people often eat food they canned themselves. Everyone has more than enough credits to meet their needs, but it's up to them what they eat. Something similar works with the other stuff too.'

They finished their meal and dropped the dirty dishes off by the serving line. 'Off to my shift', said Carmen, giving a stiff salute. 'Thomas, you're off, right? Can you take William back to the farm?'

'Sure thing', he said. Carmen and Octavia walked farther into the factory, while Thomas and Guest turned the other way and emerged into the sun.

'Longer lunch than I anticipated', said Thomas. 'Socialists love meetings.'

'Very true', Guest said with a laugh.

They biked back to the farm, and Guest was struck by how much he felt at home already. As if triggered by his thoughts, he heard someone yelling to his left. 'William!' shouted Amara, and he saw her waving. She was in a broad hat and boots, with thick gloves covering her hands, and stood among a couple dozen workers.

'I think I'll stop here', he said to Thomas. 'Nice to meet you.' The other man nodded and sped off, while Guest lumbered down the hill.

'Big mistake', said Amara after greeting him with a quick hug. 'We're putting you to work, new boy!'

The group laughed. Amara handed him boots, gloves, a hat, and a pair of overalls. 'Figured I'd see you today. Put this on and get ready to sweat', she said.

Guest obeyed, though it took him a few embarrassing minutes to wiggle into the work clothes. Amara didn't wait and was already explaining the farming methods while Guest struggled to push his boot through the bottom of the overalls.

'In this field, we're growing an ancient set of crops called the Three Sisters: maize, beans, and winter squash', Amara said. 'A clever arrangement, passed on from the region's original inhabitants, the Nipmuc, who, of course, are still here. The beans can climb up the tall corn stalks, instead of a wooden or metal pole. The maize and squash are fertilized by the nitrogen released by the beans, while the bristly squash stem keeps animals away. All increase yields and protect the health of the soil. A lot more labour, for sure, but at least we don't need chemical pesticides or fossil fuels anymore.'

She paused as she lifted up an enormous yellow flower from one of the squash vines. 'This will be a handsome butternut squash, if all goes well', she said. 'Perfect for Christmas dinner.' She tugged at a beanstalk. 'In many of the fields, we use trellises for the beans. Because a lot of our corn still comes from the industrial farms out west, we need a different ratio now than in the traditional arrangement.'

She turned to Guest. 'But you'll learn all that in school! Today, tough guy, you'll be with me on ditch duty.' They walked towards the edge of the field and saw a half-finished ditch marking a border between the crops and the woods. 'We call it "drainage management". The worst chore to be assigned. It's your lucky day, William', she said, handing him a shovel.

They went to work, along with about half of the other workers. Luckily the sun had fallen lower in the sky and was hidden behind the tops of the trees. Guest quickly fell into a rhythm, and the crunch of the shovel gave a percussive background to the murmuring conversation.

'So what's Boston like now?' he asked Amara, wiping the sweat from his eyes with his forearm. 'Half-buried and rewilded, like the farm?'

'No, not at all', she said with a laugh. 'But there is less of a difference between cities and the countryside now. Lot more workers out here these days, tending the crops. It's funny how much labour fossil fuels really saved. We still use plenty of machines, to be sure, but oil went into other things too, like fertilizer. Muscle power has to make up the gap.' She flexed her arms and laughed.

'And plenty of city workers were in industries that don't matter much anymore, like advertisement or gig economy delivery people', she added. 'Boston's still a vibrant place, and the regional hub for New England's planners and councils, but I think it's nice to have these smaller towns and communities come back to life.'

They worked for a while. 'It's much more beautiful now', Amara said, emerging from deep within her thoughts. 'Boston, I mean. There are salmon and great blue herons in the Charles River now, and the parking lots and golf courses are now gardens or rewilded ecosystems. Many of the buildings look different too. The planners have been pushing a programme to make passive homes for everyone, almost entirely by expanding, adapting, and retrofitting old buildings. Like the dormitories, super-insulated and everything. All the cold regions of the world are pushing something similar. But it's slow work.'

'Are most of the people here from Boston?' Guest asked.

'Some of us are, and some of us are from the Connecticut River Valley', said Amara. 'A fair number on the farm are climate refugees of one type or another. Carmen came here when they were a baby, after a horrible storm set off landslides in Guatemala. The twenties and thirties were a hard time for the planet. Lots of disasters.' She shook her head. 'Hopefully the worst is over, but the planners still assume that there will be bad fires and storms every year. Probably will be for many years, because unbuilding takes a long time. Were there problems where you were living?'

'Huh?' said Guest, before remembering that he said he had been abroad. 'Oh no, nothing too crazy', he said hurriedly. Amara smiled, and they returned to the ditch.

About an hour in, Guest found himself slowing down, his shoulders and back exhausted from the digging. 'Don't worry, you'll get used to it eventually', said Amara, laughing. 'Plus, since we got the hard shift, we can quit before the others.'

Guest pushed through another quarter hour before he had to throw in the towel. 'Here, take the dirt over to the side of the fields', said Amara, pointing to a wheelbarrow. Soon, he fell into a pleasant routine, moving piles of earth and chatting leisurely with other workers on the way. The hours fell away.

Eventually, Amara grabbed his arm. 'Time for dinner', she said.

Guest was surprised the rest of the shift had passed so quickly. They biked back to the dormitory in silence, tools pulled behind them in a rattling cart, enjoying the smells and sounds of the evening. As the cool air whipped through his hair, Guest thought he heard a howling in the distance.

'Are those—'

'Yes, wolves', said Amara. 'It took a long time, but they're back. They've done wonders for the ecosystem. We were overrun with deer, which disrupted plant life cycles and carried all sorts of nasty stuff like Lyme disease. This place was meant to have an apex predator. Crazy we ever drove them away. He-Yin must be proud.'

They walked into the dormitory, and Amara led him to the dining room. Much more welcoming than the solar panel factory's, he thought. The circular tables were draped with table-cloths with woven rose patterns. Morris' work, he was sure.

'William!'

He turned and saw Edith, sitting at a table with two young families. He joined them, recollecting his day and listening to theirs. Eventually he stood up and walked to the serving line. The food was delicious, a bit more indulgent than his sinless lunch. He licked his fingers from the crispy fried squash and bit into an enormous black bean burger, bursting with lettuce and tomato. The sesame bun must have been fresh from the oven; it was soft and yeasty, with a surprisingly crisp crust. The beer was dark and malty, but not too heavy. 'He-Yin's work', said Amara as she walked by, flanked by a small entourage of friends.

Guest mostly listened as his table talked about the local baseball league, music, and the next painting project for the dorms. The kids were especially excited about being able to draw on the back stairwell and talked among themselves about their design schemes. The conversation went on for hours, and some tables pulled out cards and games. Guest felt exhaustion setting in and excused himself from the table.

'We've already set you up in the room you slept in last night', said Edith. 'You'll report for classes first thing tomorrow morning.'

Guest walked to his room and saw that it had been decorated with colourful welcome signs. He was surprised by his tears; he wiped them away with the back of his hand. As he lay beneath Morris' beautiful quilt, Guest felt his heart grow heavy. The next morning, he knew, he would awaken back home, in the world as he'd left it. But as he thought back to his conversation in the bar the night before, he felt something like hope start to grow. 'If others can see it as I have seen it', he said to himself, in something like a prayer, 'then it may be called a vision rather than a dream.'

Epilogue

An Epoch of Rest

> We live in capitalism. Its power seems inescapable. So did the
> divine right of kings.
>
> —Ursula K. Le Guin

William Morris, one of the wealthiest men in Victorian
England, spent his life translating Viking sagas, weaving,
writing, and fighting for a socialist revolution. In 1890, he
published the novel *News from Nowhere*, an early eco-socialist
utopia. It tells of a disillusioned Victorian socialist, William
Guest, who magically wakes up in London in the year 2102.
He explores the paradise of socialist England, where people are
healthy, carefree, and content with few but beautiful posses-
sions. The clear Thames teems with salmon. A denizen of this
utopia tells Guest, 'The spirit of the new days, of our days, was
to be delight in the life of the world; intense and overweening
love of the very skin and surface of the earth on which man
dwells.'[1]

Like Neurath, Morris recognized that utopia was hardly an
impractical romance to be discarded by tough-minded revolutionaries, but rather a necessary part of the practical work
to realize socialism. Morris wrote his novel as a rebuttal to
Edward Bellamy's *Looking Backward*, allowing the two to
debate the future of socialism as social engineers. It was only
through utopian fiction that they could simultaneously discuss
the purpose of labour, the interchange between society and
nature, the relationship between men and women, and how
socialists could wrest power from an unyielding capitalist elite.

Much has changed since Morris dreamed of an eco-socialist future, but the need for a vibrant utopian socialism has not lessened. We have written *Half-Earth Socialism* as a humble contribution to this mighty but lately neglected tradition. Morris used his dexterous hands, his facility with languages, and his tireless dedication as an organizer to understand and build socialism for his time. We have used epidemiology, history, mathematics, philosophy, fiction – and even a nifty website – to strive towards the same end.

Our book draws on these different disciplines to understand the riddle of socialism in an age of environmental catastrophe. As we saw in chapter one, humanity faces a choice: to futilely try to further the humanization of nature through mad schemes like geoengineering, or to plan an economy within planetary boundaries. Chapter two outlined the shortcomings of the three demi-utopian solutions proposed by mainstream environmentalists: BECCS, nuclear power, and a colonial Half-Earth. We then propose alternatives for unbuilding the world, such as widespread veganism and energy quotas. However, it is the third chapter that tackles the difficult problem of organizing production and distribution without the market. Half-Earth socialist planning is inspired by a slew of traditions, including Neurath's in natura calculation, Kantorovich's linear programming, Beer's Cybersyn, and the climate-economy models of Austria's IIASA. By combining the strengths of both democratic and flexible centralized planning, our scheme aims to avert the humanitarian and ecological catastrophes of past socialist experiments. With climate disasters, biodiversity collapse, and geoengineering looming on the horizon, it has never been more important for socialists to act as what the critic Walter Benjamin called the 'emergency brake' on the locomotive of history. Socialism offers the only path to human freedom and harmony with nature, but it is no automatic utopia. To represent what Half-Earth socialism might look like in practice, we tried our hand at fiction in chapter four, in a sort of twenty-first century *News from Nowhere*. Our utopia is a modest idyll. Half-Earth socialists do not 'lead lives equivalent ... to those of today's billionaires', as the accelerationist Left promises.[2] Instead, they acknowledge that there is no escape from trade-offs between luxury and environmental stability. They can see that true freedom lies in the comprehension and acceptance of such limits.

Like the conditions imagined by Morris in *News from Nowhere*, the living standards in Half-Earth socialism only appear austere if one compares them with fantasies of abundance wrested from a fully humanized nature. Half-Earth socialism would be neither technophobic nor technophilic; rather, the adoption of advanced technologies to further the appropriation of nature would be measured against their in natura costs. We expect our utopians to enjoy more comfort than the Cubans during the Período Especial, but less than a typical eco-yuppie in the Global North who installs solar panels on the roof of their McMansion, barbecues grass-fed beef hamburgers, and has a $70,000 electric car in the garage. Half-Earth socialism would provide everyone with the essentials from health care to childcare, but occasionally it might be necessary to stand in queue. No one will be above manual work – brain surgeons and master planners will tend gardens and clean communal kitchens too.

We hope that after reading this book you will agree that some sacrifices are justified to avoid the worst of the environmental crisis. This crisis discredits the Promethean strand of socialist thought that continues to exalt the humanization of nature as a source of liberation rather than unknowable danger. Neoliberals blithely risk devastation to nature and society alike in order to protect the sacred market from the grubby control of mere mortals. Yet the Biosphere 2 fiasco demonstrated that a natural climate and a stable biosphere not only are irreplaceable preconditions for survival but are vastly more complex than we will ever know and can never be controlled. Half-Earth socialism would maintain and enhance the biosphere through abolishing animal husbandry, rebuilding cities, and rewilding at least half the planet. In such a future, we would have equality, leisure, health, and economic democracy – all utopian achievements worth fighting for in themselves rather than forced upon us by the environmental crisis. Consumerism is the golden shackle that must be cut to achieve true freedom.

The subtitle of *News from Nowhere* is a curious one: *An Epoch of Rest*. Indeed, the characters are often taking a break or extolling the virtues of rest. For Morris, rest is something more profound than mere idleness.[3] In true Hegelian fashion (despite having never read the old Swabian), Morris imagined the end of history as an era brought about by universal satisfaction, a time when all the work is done. The importance of rest is why Morris so vehemently opposed Bellamy's hyper-mechanized and consumerist 'cockney paradise'.[4] 'The development of man's resources,' Morris wrote in his review of *Looking Backward*, 'which has given him greater power over nature, has driven him also into fresh desires and fresh demands on nature … and the multiplication of machinery will just – multiply machinery'.[5] Elsewhere, Morris imagined that in socialism, once people embraced leisure over labour, then 'Nature, relieved by the relaxation of man's work, would be recovering her ancient beauty'.[6] While Hegel and Bellamy acclaimed the humanization of nature, Morris understood that eventually people would need to set down their tools and leave nature to itself.

Return of the Utopian Tradition

Perhaps the best proof of the emancipatory power of utopianism is how much the neoliberals despise it. Although they themselves are a strange breed of anti-humanist utopians who believe with a mystic's fervour in the omniscience of their Market-Moloch, they see themselves as utopia slayers. After all, neoliberalism emerged from Mises' and Hayek's engagement with Neurath – a utopian par excellence – who proposed a system of socialist governance with ambitious clarity. Midway through the socialist calculation debate, Hayek declared the purpose of economics to be the 'refutation of successive Utopian proposals'.[7]

His epistemic critique of socialism in the 1930s and 1940s was part of this effort, but he also delved into intellectual history. His *Counter-Revolution in Science* is a disparate genealogy linking Descartes, Comte, Saint-Simon, Hegel, Marx, and Keynes. They were all enemies of neoliberalism because, whatever the differences among them, they shared the belief that conscious control over humanity's destiny was possible. 'Whether it is the strict followers of Hegel who adumbrate the master's view of Reason becoming conscious of itself and taking control of its fate,' Hayek summarized, 'or Dr. Karl Mannheim [author of *Ideology and Utopia*], who thinks that "man's thought has become more spontaneous and absolute than it ever was, since it now perceives the possibility of determining itself", the basic attitude is the same.'[8] The neoliberals' utopia of the all-knowing market depends upon vanquishing all other utopian endeavours. We try to parse the differences between intellectual traditions more carefully than Hayek, especially when it comes to the three main currents of environmental thought since 1798.

By doing so, we can see some of the differences between utopian and Promethean socialists. Without a direct lineage from Hegel and his humanization of nature, the utopian socialists could believe in the perfectibility of society without assuming that it required hyper-mechanization or the domination of nature. This is perhaps why so many (but certainly not all) cared about nature and were vegetarian. Delineating this genealogy gives Half-Earth socialism a greater historical depth, tying it to liberatory traditions since the eighteenth century and arguably earlier to More and Plato. Knowing of the rival utopian socialist lineage may also help environmentalists and animal-rights activists feel less estranged from a Left that has been long dominated by Prometheans. John Oswald, a Scottish Jacobin in the thick of the French Revolution, believed in the congruence of political revolution and a revolution in humanity's relation to nature: 'the day is beginning to approach when the growing sentiment of peace and good-will towards men will also embrace, in a wide circle of benevolence, the lower orders of life.'[9]

British socialist utopians in this period were a tightly knit group. Oswald might have been an acquaintance of William Godwin, whose utopian writings provoked Malthus to write his *Essay on Population*.[10] Godwin, who dabbled in vegetarianism, knew herbivorous radicals Robert Owen and Percy Shelley. Owen organized utopian settlements in New Lanark in Scotland and New Harmony in Indiana, where he fostered innovative childhood education programmes and workers' co-operatives. Shelley – a poet whom Marx considered 'one of the advanced guard of socialism' – not only admired Oswald's pamphlet but authored the vegetarian essay *A Vindication of Natural Diet*. There, he made the Platonic argument that 'devouring an acre at a meal' led to war, and included the Jennerite insight that disease itself was historical and 'flowed from unnatural diet', which included meat.[11] He was married to Mary Shelley, the daughter of Godwin and Mary Wollstonecraft.[12]

Although it is often overlooked, Mary Shelley's *Frankenstein: Or the Modern Prometheus* grapples repeatedly with the problem of vegetarianism. The 'monster' chooses not to 'destroy the lamb and the kid, to glut my appetite; acorns and berries afford me sufficient nourishment'. Critics such as Carl J. Adams and Timothy Morton note the importance of the Prometheus myth for the Shelleys because they believed it described the prehistoric rupture when fire was used to cook meat, thus ending the Edenic vegetarian era of human history. The monster is a 'modern Prometheus' not only because he is a 'new man' born of the mangled French Revolution, but also because he *chooses* not to eat meat.[13] Notably, Shelley was inspired by François-Félix Nogaret's *Le Miroir des événemens actuels* (The mirror of recent events), a novel written during the early optimistic days of the French Revolution, in which an inventor named 'Frankénsteïn' creates a perfect mechanical automaton.[14] By contrast, Shelley's monster is made of sewn flesh stolen from proletarian corpses, and Frankenstein's desire to create a 'new species' by instilling human consciousness in inert nature ends in death and chaos.[15]

The link between utopian socialism and vegetarianism lasted deep into the nineteenth century, but it became increasingly threatened by the fire of Prometheanism. Morris himself was not a vegetarian, but his friend Edward Carpenter was. Carpenter has been called the 'gay godfather of the British Left', given his commitment to anti-vivisectionism, socialism, and sandals.[16] Vegetarianism was rife in the suffragist movement, especially among its more radical socialist elements.[17] Rank and file militants were vegetarian, as were some of the movement's better-known leaders including Isabella Ford and Charlotte Despard, as well as Charlotte Perkins Gilman in the US.

Marx, who had been living in England since 1849, appeared little affected by the tradition of British utopian socialism. Already in the 1860s, he dismissed the Royal Society for the Prevention of Cruelty to Animals for offering little more than the 'odour of sanctity' to bourgeois oppression.[18] By the twentieth century, the utopian socialist tradition was in retreat. George Orwell – every conservative's favourite socialist – excoriated vegetarians and sandals to expunge the Left of effeminate ridiculousness.[19] Utopian socialism *is* weird – but that weirdness is a source of strength. We hope that Orwell spins in his grave when Half-Earth socialism becomes a movement, radiant in its queer, feminist, vegetarian glory.

The utopian socialist tradition never went away, even as Promethean Marxism became dominant over the last century. That Neurath was such a creative thinker but exerted relatively little influence on the broader Left during the mid-twentieth century reveals both the strength of utopian socialist thought and its increasing isolation from social movements. The other inheritor of the unorthodox Left was the Frankfurt School, whose theorists criticized the destruction wrought by the mindless conquest of nature. Adorno, Horkheimer, and Benjamin, as well as Alfred Schmidt and Herbert Marcuse, all contributed to the foundations of eco-socialism. Marcuse's most famous student, Angela Davis, is not only a tireless advocate of the great utopian project of today – prison abolition – but also a vegan. The point, however, is not simply to substitute socialist utopianism for Promethean Marxism, but rather to strive for a synthesis of the two to create a new, epistemically humble socialism.

Dark Places

Today, the most prominent inheritors of utopian thinking are science fiction writers – those children of Mary Shelley. Ursula Le Guin's best-known utopian novel is likely *The Dispossessed*, which examines capitalism, state socialism, and anarchism in the planetary system of Tau Ceti, yet the lessons of Jennerite socialism might be best expressed in her novel *A Wizard of Earthsea*. Like Morris' *News from Nowhere*, it describes a society predicated on a humble and incomplete domination of nature. Everything in Earthsea is divided between surface and essence, a boundary that only magic can overcome. Knowledge of the Old Speech, the language of 'true names' that belong to nature and humans, is the source of magic in Earthsea. The book's protagonist, Sparrowhawk – whose true name is Ged – must learn 'the name of every cape, point, bay, sound, inlet, channel, harbor, shallows, reef and rock of the shores of

Lossow, a little islet of the Pelnish Sea.' Yet the wizards' knowledge of the Old Speech is forever incomplete. 'The lists are not finished', Ged is told. 'Nor will they be, till world's end.'[20] Similarly, our knowledge of nature will always be incomplete. Ged is warned that wizards should try to control only what they 'can name exactly and wholly', for it is dangerous to use the true name of something they do not understand. Like Frankenstein, hubris drives Ged to try to resurrect the dead. Because he only partially understands the spell, a 'shadow' leaps through the opening created by the spell and attacks him. Our shadows are zoonoses, climate change, the hole in the ozone layer, and unknown future horrors as the humanization of nature hurtles blindly forward.

If Ged's desire for glory and domination leads to catastrophe, Half-Earth socialism errs on the side of humility and restraint. The state is still political – conscious control over the economy means the politicization of everything – but once equality is achieved through the universal recognition of each other's worth and the natural world is stabilized, then statecraft will be reduced to mere administration. It will 'wither away' in the sense that politics will eventually be as mundane as devising bus routes. Instead of succumbing to fervid nationalist passions, the international socialist can relate to broader humanity as if they were fellow bus passengers. Care, curiosity, and self-awareness will surpass domination over fellow humans and the natural world during this epoch of rest. In this way, Le Guin's humble utopianism is similar to the works of Morris, More, and Plato. In her books, observes critic Colin Burrow, 'civilization isn't about conquering planets or travelling faster than the speed of light. It's about keeping going even when you think you're lost, recognizing that living means keeping children alive, growing fruit trees, watching things change and tending the goats.'[21] Civilization is about *life*.

Some might fault our utopia for lacking a certain grandiosity. Why not build thousands of nuclear reactors or control the climate like a thermostat? Or perhaps master closed systems like Biosphere 2 and then colonize space? We counter that creating Half-Earth and ensuring the good life for billions of people will be difficult enough tasks to occupy us for decades, even centuries. The stars can wait.

Le Guin was not drawn to old-fashioned science fiction with its visions of technological splendour. Instead, she focused on stories about people learning the lesson of humility. She talked of failure, of despair, of 'dark places'. Humanity is in a dark place now, with catastrophe after catastrophe piling one upon the other. Le Guin counsels that 'our roots are in the dark; the earth is our country ... what hope we have lies there.'[22] This is why we turn to Half-Earth, the humble endeavour of unbuilding the world. Yet it is a task as beautiful as the world we live in and feasible once our resourceful species is liberated from capital. What happens after that, who knows. As Le Guin puts it:

> there is room enough to keep even Man where he belongs, in his place in the scheme of things; there is time enough to gather plenty of wild oats and sow them too ... and still the story isn't over. Still there are seeds to be gathered, and room in the bag of stars.[23]

Acknowledgements

Let this grisly beginning be none other to you than is to wayfarers a rugged and steep mountain.

–Giovanni Boccaccio

As we put the finishing touches on this book, a year and a half after our initial deadline of February 2020, it is fitting that one of us has just moved to Florence, the adopted hometown of Giovanni Boccaccio. Nearly 700 years ago, his *Decameron* depicted a group of young people who fled to nearby Fiesole to wait out the plague ravaging the city below. They passed the time spinning stories, much like we did because of our ill-fated procrastination. *Half-Earth Socialism* provided a sort of sanctuary where we could think through the chaos of the present and provide companionship in a period of bleak isolation. What emerged was a manuscript that was much more ambitious (and three times longer) than we originally envisioned, which is why we are especially appreciative of the counsel and patience of our editors Jessie Kindig, Caitlin Doherty, Mark Martin, Jeanne Tao, and Jane Halsey. We also thank the broader team at Verso for making such a beautiful book.

Many others have contributed labour time to this project, their various efforts congealed into the commodity you now hold in your hands. The unofficial third author of *Half-Earth Socialism* is Filip Mesko, who provided extensive edits on the manuscript throughout the writing process, fact-checked the book, and offered thoughtful counter-arguments to many of the ideas in the text. Our collaboration continues: we've since brought *Half-Earth Socialism* into Filip's profession of architecture. Jose Alfredo Ramirez at the Architectural Association has been a generous supporter of our foray into this field, including hosting a virtual seminar for the book. With Filip, we have also presented *Half-Earth Socialism* at Harvard University's Graduate School of Design, thanks to an invitation from William Conroy. Harvard was also our institutional home as we wrote the book, and we are especially grateful to the Weatherhead Center and the Department of Environmental Science and Engineering.

We wish to thank Lisa Borst, Sarah Gale, Dayton Martindale, and Adam Dickerson for making the book far richer thanks to their critique (sometimes of multiple drafts) and liberal use of the Track Changes function. Frank Cahill, Cameron Hu, Paige LeComer, Michael Vettese, Vignesh Sridharan, Andreas Malm, Ken Fish, Astra Taylor, Jasper Bernes, and Keith Woodhouse all provided invaluable feedback. Gregory Vettese was not only our first reader, but the muse animating our writing. Shannon Beattie offered detailed comments and endless patience through many late-night discussions. Liu Xinyue provided many welcome distractions from reading and writing. Joy Simms patiently discussed many of the ideas in this book for years. Harvey Locke was kind enough to chat with us about the WILD Foundation and, though he might not agree with our history of Half-Earth, we hope he does not find it unfair.

We would also like to thank the community in the history department at the University of Chicago, especially Charlotte Robinson, Nicholas O'Neill, and Fredrik Albritton Jonsson, for hosting the first presentation of our ideas in October 2020. Their critique of this early draft of the book was extraordinarily useful, while William Sewell even read the third chapter twice and gave us much-needed words of encouragement when our ideas were still quite rough. Aaron Benanav, a fellow fan of Neurath, offered good humour and much insight into socialist planning and utopianism. Oliver Cussen was the first to teach

Half-Earth socialism, and we thank him and his students for an excellent discussion.

The *Half-Earth Socialism* video game, available at http://half.earth, was programmed and designed by the talented Francis Tseng and Son La Pham. Chiara Di Leone led our cybernetic reading group, which shaped the ideas behind the game and its interface. We would also like to thank Arthur Röing Baer and the entire team at Trust for their support in creating such a gorgeous website.

Finally, we would like to thank our parents for their support – their love is a haven in these Decameronian times. Fred Vettese helped with his eagle-eyed editing. Sharolyn Mathieu Vettese offered many insights and the hospitality of the 'baby Banff' writing retreat, where chapters one and two were drafted. In addition to their unwavering love and support, Tom and Susan Pendergrass offered refuge after an early COVID eviction.

Friedrich Engels once mocked William Morris as a 'sentimental socialist', an insult Morris wore as a badge of honour ('I *am* a sentimentalist … and I am proud of the title'). Allow us a moment of sentimentality at the close of this book. Without the support of all these individuals, many of whom we only met because of this sprawling project, *Half-Earth Socialism* would remain just an idea. Of course, the remaining mistakes are our responsibility (and likely one of these amazing people warned us). Admittedly, writing a book and changing the world are collaborative efforts of different orders of magnitude, but the help and inspiration we received when working on *Half-Earth Socialism* give us some hope that real change is possible. We just wish it comes before the SRM-cooled sky or a new zoonotic pandemic forces us to again retreat like Boccaccio and Shelley into the shelter of the written word. This is because we want to believe in utopias, not the horror story of a future without a summer.

Appendix

Assumptions for the linear programming model outlined in chapter three are as follows.

We assume a population of 10 billion people, all to be supplied a nutritionally sufficient diet and allotted an equal energy quota. Total habitable land area is 104,000,000 km² (see Hannah Ritchie and Max Roser, 'Land Use', *Our World in Data*, September 2019, ourworldindata.org).

We assume two biophysical boundaries: at least 50 per cent of land (52,000,000 km²) must be left to nature, and carbon emissions must be mitigated such that, assuming moderate climate sensitivity, temperatures will not exceed a small amount of warming (1.5 or 2°C, depending on the plan).

For a 67 per cent chance of keeping warming below 1.5°C, we assume a remaining carbon budget of 570 gigatonnes of CO_2, generous given that four years have passed since the IPCC estimate was made. Assuming equal emissions across the population for the remainder of the century, that gives 0.73 tonnes CO_2/year/person. Limiting warming to 2°C is much easier, with a budget of 1,320 gigatonnes CO_2, corresponding to 1.69 tonnes CO_2/year/person. Although rewilding and even some deployment of carbon-removal technologies would reduce CO_2 and give some wiggle room, we do not factor this into our calculations. (For these carbon budgets, see Kelly Levin, 'According to New IPCC Report, the World Is on Track to Exceed Its "Carbon Budget" in 12 Years', World Resources Institute, 7 October 2018, wri.org.)

Land use and emissions for various diets were mentioned

earlier in chapter three, but to reiterate, there are three options the linear programming algorithm can choose from: omnivory, with 1.08 hectares and 2.05 tonnes carbon/year/person; vegetarianism, with 0.14 hectares and 1.39 tonnes; and veganism, with 0.13 hectares and 1.05 tonnes. The sum of vegetarians, vegans, and omnivores equals the global population.

We assume some portion of the world's agriculture is 'regenerative', meaning that reforms are taken to protect soil carbon and reduce emissions. While some regenerative-agriculture advocates think emissions could be negative, we take the more modest figure of 70 per cent reduction cited earlier. We assume regenerative agriculture has lower yields and therefore needs more land to produce as much food as conventional agriculture; following an estimate for organic crops, we estimate a 34 per cent reduction in regenerative yields. (See Verena Seufert et al., 'Comparing the Yields of Organic and Conventional Agriculture', *Nature* 485, no. 7397 [2012]: 229–32.) We pick an ambitious 80 per cent of agriculture to be 'regenerative' in this sense, though as long as agricultural emissions are reduced below the per capita allowance (most relevant in the 1.5°C case), model output isn't overly sensitive to this quantity.

Emissions and land-use figures for different power generation methods are assumed as follows: coal and petroleum have emissions of 8.5 and 8.8 kg CO_2/year/W respectively, calculated using the same source as the methane (natural gas) emissions in the text, while their land-use costs are 1,000 W/m^2 for coal (high variance as with methane; Smil, *Power Density*, 140) and 650 W/m^2 for petroleum (2012 global mean for extraction; Smil, *Power Density*, 115); methane, to reiterate the text, costs 3.6 kg CO_2/year/W with a power density of 4,500 W/m^2; biofuels, concentrated solar power (CSP), photovoltaics, and wind are all assumed to have zero emissions; CSP, photovoltaics, and wind have power densities of 20, 10, and 50 W/m^2 respectively, while liquid biofuels from a mix of ethanol and biodiesel have power densities of 0.3 W/m^2, and solid biofuels from wood and gaseous biofuels from phytomass have a density of 0.9 W/m^2 (Box 8.1 in Smil, *Power Density*, 246).

Following Smil's estimates, we assume wind and solar power compose equal parts of electricity generation. In the US in 2012, 15 per cent of fossil-fuel use went to electricity, 52 per cent to liquid fuels, and 33 per cent to solid and gaseous fuels. Hence we consider four scenarios: one where just electricity generation is made renewable but the overall mix remains the same; one where electricity generation is renewable and fuel use falls by 50 per cent due to reforms such as the strict rationing of transportation, an uptick in recycling materials like steel, and curtailment of other resource use (new energy mix is 26 per cent electricity, 45 per cent liquid fuels, 29 per cent solid/gaseous fuels); one where all but 10 per cent of liquid fuels are entirely electrified and solid/gaseous fuel use halved (new mix 74 per cent electricity, 6 per cent liquid fuels, and 20 per cent solid/gaseous fuels); and one of total electrification of all sectors. For electricity generation, the linear programming model is free to choose any fuel source. For liquid fuels, it must meet the quota using one of liquid biofuels or petroleum. For solid/gaseous fuels, it must use solid/gaseous biofuels, methane, or coal.

The model is run using the PuLP package in Python.

Notes

Introduction

1 Thomas Knutson et al., 'Tropical Cyclones and Climate Change Assessment: Part II: Projected Response to Anthropogenic Warming', *Bulletin of the American Meteorological Society* 101, no. 3 (2020): E303–22.

2 David Keith, *A Case for Climate Geoengineering* (MIT Press, 2013), 71; David Keith, 'The Perils and Promise of Solar Geoengineering', lecture at Harvard Museum of Natural History, Cambridge, MA, 30 October 2019, YouTube video, 1:05:28, youtube.com.

3 David W. Keith et al., 'Stratospheric Solar Geoengineering Without Ozone Loss', *Proceedings of the National Academy of Sciences* 113, no. 52 (2016): 14910–14.

4 Daniel J. Cziczo et al., 'Unanticipated Side Effects of Stratospheric Albedo Modification Proposals Due to Aerosol Composition and Phase', *Scientific Reports* 9, no. 18825 (2019).

5 'MOP 30 Highlights', *Earth Negotiations Bulletin* 19, no. 144 (2018), enb.iisd.org; Sara Stefanini, 'US and Saudi Arabia Block Geoengineering Governance Push', *Climate Home News*, 14 March 2019, climatechangenews.com.

6 Joshua Horton and David Keith, 'Solar Geoengineering and Obligations to the Global Poor', in *Climate Justice and Geoengineering: Ethics and Policy in the Atmospheric Anthropocene*, ed. Christopher J. Preston (Rowman & Littlefield, 2016).

7 John Lauerman and Tasos Vossos, 'Pandemic Bonds Paying 11% Face Their Limits in Ebola-Hit Congo', *Bloomberg Quint*, 14 August 2019, bloombergquint.com.

8 Jeremy Blackman, Micah Maidenberg, and Sylvia Varnham O'Regan, 'Mexico's Disaster Bonds Were Meant to Provide

Quick Cash after Hurricanes and Earthquakes. But It Often Hasn't Worked Out That Way', *Los Angeles Times*, 8 April 2018, latimes.com.

9 Christopher J. Smith et al., 'Impacts of Stratospheric Sulfate Geoengineering on Global Solar Photovoltaic and Concentrating Solar Power Resource', *Journal of Applied Meteorology and Climatology* 56, no. 5 (2017): 1483–97; ExxonMobil, *Outlook for Energy: A Perspective to 2040* (2019), 30, corporate. exxonmobil.com/Energy-and-environment/Looking-forward/ Outlook-for-Energy.

10 'The World Urgently Needs to Expand Its Use of Carbon Prices', *Economist*, 23 May 2020, economist.com.

11 To keep warming below 1.5°C, the IPCC estimates that carbon taxes will have to fall into the range of $135 to $6,050 by 2030. Marc Hafstead and Paul Picciano, 'Calculating Various Fuel Prices under a Carbon Tax', *Resources*, 28 November 2017, resourcesmag.org; IPCC, *Global Warming of 1.5°C: An IPCC Special Report* (2018), 152.

12 ExxonMobil, *Outlook for Energy*, 9.

13 Ibid., 14.

14 Ibid., 28.

15 'A World Turned Upside Down', *Economist*, 25 February 2017; Jacques Leslie, 'Utilities Grapple with Rooftop Solar and the New Energy Landscape', *Yale Environment 360*, 31 August 2017, e360.yale.edu.

16 There are no good options for storage in a renewable energy system. Dams are quite destructive, while batteries have a very low energy density. Richard Heinberg and David Fridley, *Our Renewable Future: Laying the Path for One Hundred Percent Clean Energy* (Island Press, 2016), 56–7.

17 David McDermott Hughes, 'To Save the Climate, Give Up the Demand for Constant Electricity', *Boston Review*, 1 October 2020, bostonreview.net.

18 Linnea I. Laestadius et al., 'No Meat, Less Meat, or Better Meat: Understanding NGO Messaging Choices Intended to Alter Meat Consumption in Light of Climate Change', *Environmental Communication* 10, no. 1 (2016): 84–103; Brian Machovina, Kenneth J. Feeley, and William J. Ripple, 'Biodiversity Conservation: The Key Is Reducing Meat Consumption', *Science of the Total Environment* 536 (2015): 419–31.

19 'Meat and Meat Products', Food and Agriculture Organization of the United Nations, fao.org/ag.

20 Mary J. Gilchrist et al., 'The Potential Role of Concentrated Animal Feeding Operations in Infectious Disease Epidemics and Antibiotic Resistance', *Environmental Health Perspectives* 115, no. 2 (2007): 313–16; Kate Kelland, 'French-German *E. coli* Link Seen in Sprouted Seeds', Reuters, 27 June 2011, reuters.com; Wim van der Hoek et al., 'Epidemic Q Fever in Humans in the Netherlands', *Advances in Experimental Medicine and Biology* 984 (2012): 329–64.

21 'Influenza: H5N1', World Health Organization, who.int/newsroom/q-a-detail/influenza-h5n1; James Sturcke, 'Burning Issue', *Guardian*, 22 August 2005.

22 For a criticism of culling, see Susanne H. Sokolow et al., 'Ecological Interventions to Prevent and Manage Zoonotic Pathogen Spillover', *Philosophical Transactions of the Royal Society B: Biological Sciences* 374, no. 1782 (2019): 20180342; Daniel G. Streicker et al., 'Ecological and Anthropogenic Drivers of Rabies Exposure in Vampire Bats: Implications for Transmission and Control', *Proceedings of the Royal Society B: Biological Sciences* 279, no. 1742 (2012): 3384–92.

23 Jonathan A. Patz et al., 'Unhealthy Landscapes: Policy Recommendations on Land Use Change and Infectious Disease Emergence', *Environmental Health Perspectives* 112, no. 10 (2004): 1092–8.

24 World Health Organization, *A Billion Voices: Listening and Responding to the Health Needs of Slum Dwellers and Informal Settlers in New Urban Settings* (WHO Kobe Centre, 2005), 4, who.int/social_determinants. For a countervailing analysis on this question, see Aaron Benanav, 'Demography and Dispossession: Explaining the Growth of the Global Informal Workforce, 1950–2000', *Social Science History* 43, no. 4 (2019): 679–703.

25 Cornelia Daheim and Ole Wintermann, *2050: The Future of Work* (Bertelsmann Stiftung, 2016), 11, bertelsmann-stiftung.de.

26 World Inequality Lab, *World Inequality Report 2018: Executive Summary*, Fig. E9, wir2018.wid.world/files.

27 Fiona Harvey, 'World's Richest 1% Cause Double CO_2 Emissions of Poorest 50%, Says Oxfam', *Guardian*, 21 September 2020.

28 Among others, David Keith, a leading geoengineer, has received death threats from chemtrails conspiracy theorists. Virginia Gewin, 'Real-Life Stories of Online Harassment – and How

Scientists Got Through It', *Nature*, 16 October 2018, nature .com.

29 Raymond Pierrehumbert, 'The Trouble with Geoengineers "Hacking the Planet"', *Bulletin of the Atomic Scientists*, 23 June 2017, thebulletin.org.

30 Simon Factor, 'The Experimental Economy of Geoengineering', *Journal of Cultural Economy* 8, no. 3 (2015): 309–24.

31 Steven D. Levitt and Stephen J. Dubner, *SuperFreakonomics: Global Cooling, Patriotic Prostitutes, and Why Suicide Bombers Should Buy Life Insurance* (William Morrow, 2011); Philip Mirowski, *Never Let a Serious Crisis Go to Waste: How Neoliberalism Survived the Financial Meltdown* (Verso, 2014), 340. Geoengineer David Keith's start-up Carbon Engineering is supported by Chevron and Murray Edwards, a Canadian tar sands magnate. See carbonengineering.com/our-team.

32 Quite a few well-known intellectuals attended this luxury-hotel liberal jamboree, including Karl Popper, Raymond Aron, Milton Friedman, Ludwig von Mises, Michael Polanyi, and Bertrand de Jouvenel. Only one woman, historian Veronica Wedgwood, was invited. See Bruce Caldwell, 'Mont Pèlerin 1947', in *From the Past to the Future: Ideas and Actions for a Free Society*, ed. John B. Taylor (Mont Pèlerin Society, 2020), 44, hoover.org.

33 Friedrich Hayek, 'The Use of Knowledge in Society', *American Economic Review* 35, no. 4 (1945): 526. Philip Mirowski summarizes Hayek's innovation as recasting the market as a sort of all-knowing 'information-processor'. Mirowski, *Never Let a Serious Crisis Go to Waste*, 54.

34 Mirowski, *Never Let a Serious Crisis Go to Waste*, 77.

35 Ibid., 332. Mirowski observes that neoliberals 'have been able to promote and coordinate interlocking full-spectrum braces of alternative policies that expand until they entirely fill the public space of perceived alternatives.' There are also many ways to make a market: see Philip Mirowski and Edward Nik-Khah, *The Knowledge We Have Lost in Information: The History of Information in Modern Economics* (Oxford University Press, 2017).

36 Stuart Hall, 'Thatcher's Lessons', *Marxism Today* (March 1988): 20.

37 Jeremy Hance, 'Could We Set Aside Half the Earth for Nature?', *Guardian*, 15 June 2016.

38 In 1967, Wilson and MacArthur offered the formula of $S = CA^z$, 'where S is the number of species, A is the area, C is a

constant that varies widely among taxa and according to the unit of area measurement, and z is a constant which falls in most cases between 0.20 and 0.35.' Edward O. Wilson and Robert H. MacArthur, *The Theory of Island Biogeography* (Princeton University Press, 1967), 17. In his recent book *Half-Earth*, Wilson uses the formula of the fourth root, equivalent to setting $z = 0.25$. Edward O. Wilson, *Half-Earth: Our Planet's Fight for Life* (Liveright, 2016), 186.

39 Sarah Gibbens, 'Less Than 3 Percent of the Ocean Is "Highly Protected"', *National Geographic*, 25 September 2019, nationalgeographic.com; Kendall R. Jones et al., 'One-Third of Global Protected Land Is under Intense Human Pressure', *Science* 360, no. 6390 (2018): 788.

40 Wilson, *Half-Earth*, 186.

41 Kate E. Jones et al., 'Global Trends in Emerging Infectious Diseases', *Nature* 451, no. 7181 (2008): 992.

42 To cite one of many grim statistics, at least 212 environmental activists were killed in 2019, with casualties disproportionately borne by Indigenous activists in Latin America. Mélissa Godin, 'Record Number of Environmental Activists Killed in 2019', *Time*, 29 July 2020, time.com.

43 Alan Beattie and James Politi, '"I Made a Mistake," Admits Greenspan', *Financial Times*, 23 October 2008, ft.com.

44 Quinn Slobodian, 'Anti-'68ers and the Racist-Libertarian Alliance: How a Schism among Austrian School Neoliberals Helped Spawn the Alt Right', *Public Culture* 15, no. 3 (2019): 372–86; Melinda Cooper, 'The Alt-Right: Neoliberalism, Libertarianism and the Fascist Temptation', *Theory, Culture & Society* (April 2021): 1–21; Quinn Slobodian and Dieter Plehwe, 'Neoliberals Against Europe', in *Mutant Neoliberalism: Market Rule and Political Ruptures*, ed. William Callison and Zachary Manfredi (Fordham University Press, 2018), 89–111.

45 'Monthly CO_2', *CO2.earth*, co2.earth/monthly-co2.

46 Richard Betts, 'Met Office: Atmospheric CO_2 Now Hitting 50% Higher Than Pre-Industrial Levels', Carbon Brief, 16 March 2021, carbonbrief.org.

47 'More Than Half of All CO_2 Emissions Since 1751 Emitted in the Last 30 Years', *Institute for European Environmental Policy*, 29 April 2020, ieep.eu; 'World of Change: Amazon Deforestation', NASA Earth Observatory, earthobservatory.nasa.gov.

48 'Deforestation Has Slowed Down but Still Remains a Concern, New UN Report Reveals', *UN News*, 21 July 2020, news.un.org.

49 Michael Perelman, *Farming for Profit in a Hungry World: Capital and the Crisis in Agriculture* (Allanheld, Osmun, 1978). The country's butchers slaughtered 2 billion chickens in 1989, a grisly number that has increased more than fivefold three decades later. FAOSTAT, fao.org.

50 Andreas Malm, 'China as Chimney of the World: The Fossil Capital Hypothesis', *Organization & Environment* 25, no. 2 (2012): 146–77.

51 We will see some examples of this dark history in chapter two, particularly in the African context. For North America, see for example Karl Jacoby, *Crimes Against Nature: Squatters, Poachers, and the Hidden History of American Conservation* (University of California Press, 2014).

52 Friedrich Hayek, 'Planning, Science and Freedom', *Nature* 148 (1941): 580. Science and technology studies, which today is dominated by the explicitly anti-Marxist philosopher Bruno Latour and his epigones, emerged as an academic field after Soviet and British Marxist philosophers (such as J. D. Bernal, Nikolai Bukharin, and Mikhail Bakhtin) met at the 1931 International Congress of the History of Science in London. See Gary Werskey, *The Visible College: The Collective Biography of British Scientific Socialists of the 1930s* (Holt, Rinehart and Winston, 1978), chapter 5.

53 Will Kymlicka, 'Human Supremacism: Why Are Animal Rights Activists Still the "Orphans of the Left"?', *New Statesman*, 30 April 2019, newstatesman.com; Ryan Gunderson, 'Marx's Comments on Animal Welfare', *Rethinking Marxism* 23, no. 4 (2011): 543–8.

54 Amie 'Breeze' Harper, 'Dear Post-Racial White Vegans: "All Lives Matter" Is a Racial Microaggression Contributing to Our Daily Struggle with Racial Battle Fatigue', *Sistah Vegan*, 13 January 2015, sistahvegan.com; Summer Anne Burton, 'Stop Comparing Black Lives Matter to Animal Rights', *Tenderly*, 4 June 2020, medium.com/tenderlymag; for a critique of the animal-rights movement from a disability studies perspective, see Sunaura Taylor, *Beasts of Burden: Animal and Disability Liberation* (New Press, 2017); David Sztybel, 'Can the Treatment of Animals Be Compared to the Holocaust?', *Ethics and the Environment* 11, no. 1 (2006): 97–132.

55 'Why Black Americans Are More Likely to Be Vegan', BBC, 11 September 2020, bbc.com.

56 Upton Sinclair, *The Jungle* (Penguin, 1985 [1905]), 408.

57 Alyssa Battistoni, 'Living, Not Just Surviving', *Jacobin*, 15 August 2017, jacobinmag.com.

58 'It's Not Intersectional, It's DxE: An Exposé Written by DxE's Victims', *Dismantle DxE*, 16 September 2015, dismantledxe. wordpress.com; Owen Hatherley, 'A Tale of Rape Claims, Abuses of Power and the Socialist Workers Party', *Guardian*, 8 February 2013.

59 Things have become better lately; see Joanna Kerr, 'Greenpeace Apology to Inuit for Impacts of Seal Campaign', Greenpeace, 24 June 2014, greenpeace.org. For a historical perspective, see Ryan Tucker Jones, 'When Environmentalists Crossed the Strait: Subsistence Whalers, Hippies, and the Soviets', *RCC Perspectives*, no. 5 (2019): 81–8.

60 Richard Schuster et al., 'Vertebrate Biodiversity on Indigenous-Managed Lands in Australia, Brazil, and Canada Equals That in Protected Areas', *Environmental Science & Policy* 101 (2019): 1–6.

61 Mahathir bin Mohamad rightly scolded delegates during the 1992 Rio Earth Summit: 'When the rich chopped down their own forests, built their poison-belching factories, and scoured the world for cheap resources, the poor said nothing. Indeed they paid for the development of the rich. Now the rich claim a right to regulate the development of poor countries.' Mahathir bin Mohamad (address to the United Nations Conference on Environment and Development, Rio de Janeiro, Brazil, 13 June 1992), mahathir.com.

62 Friedrich Engels, 'Preface to the English Edition', in *Capital: A Critique of Political Economy*, vol. 1, by Karl Marx (Penguin, 1976 [1887]), 113.

63 Eric Hobsbawm, 'The Murder of Chile', *New Society*, 20 September 1973, quoted in Ralph Miliband, 'The Coup in Chile', *Jacobin*, 11 September 2016.

64 *Looking Backward* was one of the real blockbusters of the nineteenth century, alongside *Uncle Tom's Cabin* and *Ben-Hur*.

65 John Atkinson Hobson, 'Edward Bellamy and the Utopian Romance', *Humanitarian* 13 (1898): 180, quoted in Matthew Beaumont, introduction to *Looking Backward: 2000–1887*,

by Edward Bellamy (Oxford University Press, 2007 [1888]), xvii.

66 Karl Marx, 'Postface to the Second Edition', in *Capital: A Critique of Political Economy*, vol. 1 (Penguin, 1976 [1867]), 99. See also, e.g., Karl Marx and Friedrich Engels, *The Communist Manifesto* (Penguin, 2015 [1848]), 46–9.

67 Philip Mirowski, 'Markets Come to Bits: Evolution, Computation and Markomata in Economic Science', *Journal of Economic Behavior & Organization* 63, no. 2 (2007): 209–42.

1. Binding Prometheus

1 Peder Anker, 'The Ecological Colonization of Space', *Environmental History* 10, no. 2 (2005): 239.

2 Cyrus K. Boynton and Arthur K. Colling, 'Solid Amine CO_2 Removal System for Submarine Application', *SAE Transactions* 92 (1983): 601; Eugene A. Ramskill, 'Nuclear Submarine Habitability', *SAE Transactions* 70 (1962): 355.

3 Buckminster Fuller, *Operating Manual for Spaceship Earth* (Simon & Schuster, 1969).

4 The geodesic dome only garnered its green veneer in 1954 at the Milan Triennale (the theme that year: 'Life Between Artifact and Nature: Design and the Environmental Challenge') when Fuller won the Gran Premio award for his 13-metre-high cardboard hemisphere that was assembled on site using directions on the cardboard itself. 'Geodesic Domes', Buckminster Fuller Institute, bfi.org.

5 For an elaboration of this term, see Anker, 'Ecological Colonization', 243.

6 After the Cold War, the US opted instead to collaborate with Russia in building a vessel that would become the International Space Station. Sabine Höhler, 'The Environment as a Life Support System: The Case of Biosphere 2', *History and Technology* 26, no. 1 (2010): 48.

7 Anker, 'Ecological Colonization', 256. Incidentally, Biosphere 2 and Disney's Epcot Center in Florida both feature geodesic domes.

8 The bees were already having a hard time, because the windowpanes blocked ultraviolet light, which they needed to guide their

sight and navigation. 'Lee Pivnik at Biosphere 2', *Art Viewer*, 9 September 2017, artviewer.org.

9 For example, in some parts of south-western China today, hand-pollinators have replaced local bees killed by pesticides. Dave Goulson, 'Pollinating by Hand: Doing Bees' Work', interview by Natalie Muller, *Deutsche Welle*, 31 July 2014, dw.com.

10 Mark Nelson, 'Biosphere 2: What Really Happened?', *Dartmouth Alumni Magazine*, May–June 2018, dartmouthalumnimagazine. com; Jane Poynter, 'What Lessons Came Out of Biosphere 2?', interview by Guy Raz, NPR, 27 September 2013, npr.org.

11 Joel E. Cohen and David Tilman, 'Biosphere 2 and Biodiversity – The Lessons So Far', *Science* 274, no. 5290 (1996): 1150.

12 James K. Wetterer et al., 'Ecological Dominance by *Paratrechina longicornis* (Hymenoptera: Formicidae), an Invasive Tramp Ant, in Biosphere 2', *Florida Entomologist* 82, no. 3 (1999): 381–8.

13 Cohen and Tilman, 'Biosphere 2 and Biodiversity', 1151.

14 Kolbert's book popularized the concept more recently, but it was coined earlier. See Richard E. Leakey and Roger Lewin, *The Sixth Extinction: Patterns of Life and the Future of Humankind* (Anchor, 1996); Norman Myers, *The Sinking Ark: A New Look at the Problem of Disappearing Species* (Pergamon Press, 1979).

15 Gerardo Ceballos, Paul R. Ehrlich, and Peter H. Raven, 'Vertebrates on the Brink as Indicators of Biological Annihilation and the Sixth Mass Extinction', *Proceedings of the National Academy of Sciences* 117, no. 24 (2020): 13596 (emphasis added).

16 Robert Costanza et al., 'Changes in the Global Value of Ecosystem Services', *Global Environmental Change* 26 (2014): 152–8. Robert Costanza started this whole genre when he first put a price on nature in 1997, some $33 trillion annually. See Robert Costanza et al., 'The Value of the World's Ecosystem Services and Natural Capital', *Nature* 387, no. 6630 (1997): 253–60.

17 Oscar Wilde, 'Lady Windermere's Fan', in *Five Plays by Oscar Wilde* (Bantam Books, 1961), 42.

18 Terry Pinkard, *Hegel: A Biography* (Cambridge University Press, 2000), 24.

19 Thomas Malthus, *An Essay on the Principle of Population, as It Affects the Future Improvement of Society: With Remarks on the Speculations of Mr. Godwin, M. Condorcet, and Other Writers* (J. Johnson, 1798), 2.

20 Andrea A. Rusnock, 'Historical Context and the Roots of Jenner's

Discovery', *Human Vaccines & Immunotherapeutics* 12, no. 8 (2012): 2027. Still, Britain's uptake of the new treatment trailed behind that of revolutionary France, whose army carried out an early mass vaccination campaign as early as 1803.

21 Raymond Plant highlights the importance of this work for Hegel's subsequent philosophy; see his 'Hegel and Political Economy (Part I)', *New Left Review* 1, no. 103 (1977): 82–4.

22 Georg W. F. Hegel, *Philosophy of Nature, Being Part Two of the Encyclopaedia of the Philosophical Sciences (1830)* (Oxford University Press, 2004), 444.

23 Georg W. F. Hegel, 'The Spirit of Christianity and Its Fate', in *On Christianity: Early Theological Writings by Friedrich Hegel*, trans. T. M. Knox (Harper & Brothers, 1961), 182.

24 Hegel notes that the Greek myth of the Flood was quite different from the Jewish tradition. While neither Nimrod nor Noah managed to return to the state of nature, 'a more beautiful pair, Deucalion and Pyrrha', enjoyed a different fate, and 'after the flood in their time, invited men once again to friendship with the world, to nature, made them forget their need and their hostility in joy and pleasure, made a peace of *love*, were the progenitors of more beautiful peoples, and made their age the mother of a newborn natural life which maintained its bloom of youth.' Hegel, 'Spirit of Christianity', 184–5. Ironically, Deucalion was Prometheus' son.

25 Ibid., 183. As Hegel explains, 'life was yet so far respected that men were prohibited from eating the blood of animals because in it lay the life, the soul, of the animals.'

26 Ibid., 184.

27 Ibid., 186.

28 Plant, 'Hegel and Political Economy (Part I)', 84.

29 Godwin was a strange opponent for Malthus, for they both were critics of the French Revolution. Godwin not only opposed 1789, but was a self-declared 'enemy of revolutions'. William Petersen, 'The Malthus–Godwin Debate, Then and Now', *Demography* 8, no. 1 (1971): 16.

30 William Godwin, 'Of Avarice and Profusion', in *The Enquirer* (John Anderson Junior, 1823 [1793]), 156–7; Malthus, *An Essay on the Principle of Population*, i.

31 Malthus, *An Essay on the Principle of Population*, 14.

32 Jenner did not know of Jesty's method because Jesty never

published his results. James F. Hammarsten, William Tattersall, and J. E. Hammarsten, 'Who Discovered Smallpox Vaccination? Edward Jenner or Benjamin Jesty?', *Transactions of the American Clinical and Climatological Association* 90 (1979): 44–55.

33 Jenner's understanding was incomplete because he did not realize at first that his vaccination method did not provide life-long immunity. Murray Dworetzky, Sheldon Cohen, and David Mullin, 'Prometheus in Gloucestershire: Edward Jenner, 1749–1823', *Journal of Allergy and Clinical Immunology* 112, no. 4 (2003): 810.

34 Arthur Boylston, 'The Origins of Vaccination: No Inoculation, No Vaccination', *Journal of the Royal Society of Medicine* 106, no. 10 (2013): 396.

35 Edward Jenner, *An Inquiry into the Causes and Effects of the Variolæ vaccinæ, a Disease Discovered in Some of the Western Counties of England, Particularly Gloucestershire, and Known by the Name of the Cow Pox* (Sampson Low, 1798), 1.

36 Ibid., 2.

37 Carlton Gyles, 'One Medicine, One Health, One World', *Canadian Veterinary Journal* 57, no. 4 (2016): 345–6; Michael Francis, 'Vaccination for One Health', *International Journal of Vaccines & Vaccination* 4, no. 5 (2017), 00090.

38 In a rare case where it is mentioned, it is mocked as a 'rambling hypothesis that many human diseases were derived from animals'. Boylston, 'The Origins of Vaccination', 395.

39 Lisa Herzog, *Inventing the Market: Smith, Hegel, and Political Theory* (Oxford University Press, 2013), 59–60.

40 Malthus, *An Essay on the Principle of Population*, 16.

41 James P. Huzel, 'The Demographic Impact of the Old Poor Laws: More Reflections on Malthus', in *Malthus and His Time*, ed. Michael Turner (Palgrave Macmillan, 1986), 40–59.

42 One of the few socialist thinkers who have explored this concept is John O'Neill. See 'Science, Wonder and the Lust of the Eyes', *Journal of Applied Philosophy* 10, no. 2 (1993): 139–46.

43 Karl Marx, *Economic and Philosophic Manuscripts of 1844* (Prometheus Books, 1988 [1932]), 77.

44 Ibid., 107–8.

45 Hesiod, 'Theogony', in *Hesiod*, trans. Richmond Lattimore (University of Michigan Press, 1959), p. 153, line 510; p. 157, line 567.

46 Karl Marx and Friedrich Engels, review of *Die Religion des*

Neuen Weltalters: Versuch einer Combinatorisch-Aphoristischen Grundlegung, by Georg Friedrich Daumer (Hamburg, 1850), in *Collected Works*, vol. 10 (Lawrence & Wishart, 1978), 245, quoted in Reiner Grundmann, 'The Ecological Challenge to Marxism', *New Left Review* 1, no. 187 (1991): 110. Jenner was the first to argue that the cuckoo was a parasitic species, a debate that would last until 1921 when photographic evidence of the behaviour settled the matter.

47 Rick Kuhn, 'Marxism and Birds', 30 March 1998, sa.org.au/marxism_page/marxbird/marxbird.htm. For Stalin, see John Lewis Gaddis, *The Landscape of History: How Historians Map the Past* (Oxford University Press, 2002), 117.

48 Karl Marx, *Capital: A Critique of Political Economy*, vol. 1 (Penguin, 1976 [1867]), 639n49; Karl Marx, 'Economic Manuscript of 1861–63', in *Collected Works*, vol. 31 (Lawrence & Wishart, 1989), 345.

49 Jonathan Sperber, *Karl Marx: A Nineteenth-Century Life* (Liveright, 2013), 354.

50 Bertell Ollman, 'Marx's Vision of Communism: A Reconstruction', *Critique* 8, no. 1 (1977): 27–8.

51 Friedrich Engels, 'Outlines of a Critique of Political Economy', in *Collected Works*, vol. 3 (Lawrence & Wishart, 1975), 440.

52 Kohei Saito, *Karl Marx's Ecosocialism: Capital, Nature, and the Unfinished Critique of Political Economy* (Monthly Review Press, 2017), 229.

53 Leon Trotsky, *Literature and Revolution* (Haymarket, 2005 [1924]), 204.

54 Stephen Brain, 'Stalin's Environmentalism', *Russian Review* 69, no. 1 (2010): 93–118; John Bellamy Foster, 'Late Soviet Ecology and the Planetary Crisis', *Monthly Review*, 1 June 2015, monthlyreview.org.

55 Boris Lyubimov, *Bering Strait Dam* (US Joint Publications Research Service, 1960), 1, apps.dtic.mil.

56 Ken Caldeira and Govindasamy Bala, 'Reflecting on 50 Years of Geoengineering Research', *Earth's Future* 5, no. 1 (2017): 10.

57 Philip Micklin, 'The Aral Sea Disaster', *Annual Review of Earth and Planetary Sciences* 35 (2007): 47–72.

58 Beatrice Grabish, 'Dry Tears of the Aral', *UN Chronicle* 1 (1999): 38–44.

59 See John Bellamy Foster and Paul Burkett, *Marx and the Earth:*

An Anti-Critique (Brill, 2016). For a review that criticizes the eco-socialist literature on this point, see Andreas Malm, 'For a Fallible and Lovable Marx: Some Thoughts on the Latest Book by Foster and Burkett', *Critical Historical Studies* 4, no. 2 (2017): 267–75.

60 Alex Williams and Nick Srnicek, '#ACCELERATE MANIFESTO for an Accelerationist Politics', *Critical Legal Thinking*, 14 May 2013, criticallegalthinking.com.

61 Leigh Phillips and Michal Rozworski, 'Planning the Good Anthropocene', *Jacobin*, 15 August 2017, jacobinmag.com; Peter Frase, 'By Any Means Necessary', *Jacobin*, 15 August 2017.

62 Holly Jean Buck, *After Geoengineering: Climate Tragedy, Repair, and Restoration* (Verso, 2019), 168, 181.

63 Ibid., 173, 178.

64 See Thomas Robertson, *The Malthusian Moment: Global Population Growth and the Birth of American Environmentalism* (Rutgers University Press, 2012); Alison Bashford and Joyce E. Chaplin, *The New Worlds of Thomas Robert Malthus: Rereading the Principle of Population* (Princeton University Press, 2016).

65 Paul R. Ehrlich, *The Population Bomb* (Ballantine Books, 1968), 15.

66 Ibid., xi.

67 Garrett Hardin, 'The Tragedy of the Commons', *Science* 162, no. 3859 (1968): 1246.

68 Garrett Hardin, 'Commentary: Living on a Lifeboat', *BioScience* 24, no. 10 (1974): 561.

69 Thomas Malthus, *An Essay on the Principle of Population, or, A View of Its Past and Present Effects on Human Happiness: With an Inquiry into Our Prospects Respecting the Future Removal or Mitigation of the Evils Which It Occasions*, 2nd ed. (J. Johnson, 1803), 6.

70 Garret Hardin, 'Conspicuous Benevolence and the Population Bomb: Why Good Fences Make Good Neighbors', *Chronicles* 15, no. 10 (1991): 18–22; 'Garrett Hardin', Southern Poverty Law Center, splcenter.org.

71 For instance, the textbook *Environment and Society: A Reader*, ed. Schlottmann et al. (New York University Press, 2017), conspicuously excerpted 'The Tragedy of the Commons' – a short essay – to leave out the most odious sections, including the one cited above.

72 Mark Tran, 'David Attenborough: Trying to Tackle Famine with Bags of Flour Is "Barmy"', *Guardian*, 18 September 2013, theguardian.com; George Monbiot, 'Population Panic Lets Rich People Off the Hook for the Climate Crisis They Are Fuelling', *Guardian*, 26 August 2020, theguardian.com.

73 The permit idea was broached by the environmental economist Kenneth Boulding as a joke (though this is disputed), but it is still taken seriously by some, such as Herman Daly. See 'Ecologies of Scale', *New Left Review* 1, no. 108 (2018): 92. In 1973, the involuntary sterilization of two Black girls, Minnie and Mary Alice Relf, in Alabama brought to national attention that the federal government annually underwrote the sterilization of 100,000 to 150,000 people who otherwise would have had their welfare benefits cut. ZPG's ambivalent stance during this tragedy led to a rift between environmentalists and African American organizations. 'Relf v. Weinberger', Southern Poverty Law Center, splcenter.org.

74 Natasha Lennard, 'The El Paso Shooter Embraced Eco-Fascism. We Can't Let the Far Right Co-opt the Environmental Struggle', Intercept, 5 August 2019, theintercept.com.

75 Michael Greger, 'The Human/Animal Interface: Emergence and Resurgence of Zoonotic Infectious Diseases', *Critical Reviews in Microbiology* 33, no. 4 (2007): 243.

76 Robin Weiss and Anthony J. McMichael, 'Social and Environmental Risk Factors in the Emergence of Infectious Diseases', *Nature Medicine* 10, no. 12 (2004): S72; Kennedy Shortridge, 'Severe Acute Respiratory Syndrome and Influenza: Virus Incursions from Southern China', *American Journal of Respiratory and Critical Care Medicine* 168, no. 12 (2003): 1417; Tony McMichael, *Human Frontiers, Environments and Disease: Past Patterns, Uncertain Futures* (Cambridge University Press, 2001), 101.

77 Igor Babkin and Irina N. Babkina, 'The Origin of the Variola Virus', *Viruses* 7, no. 3 (2015): 1106–7.

78 Nathan Wolfe, Claire Panosian Dunavan, and Jared Diamond, 'Origins of Major Human Infectious Diseases', *Nature* 447, no. 7142 (2007): 282.

79 Mark S. Smolinski, Margaret A. Hamburg, and Joshua Lederberg, eds, *Microbial Threats to Health: Emergence, Detection, and Response* (National Academies Press, 2003), 17.

80 C. E. Gordon Smith, 'Introductory Remarks', in *Ebola Virus Hemorrhagic Fever*, ed. S. R. Pattyn (Elsevier, 1978), 13.

81 The author continued: 'Such a change, if sufficiently adopted or imposed, could still reduce the chances of the much-feared [avian] influenza epidemic.' David Benatar, 'The Chickens Come Home to Roost', *American Journal of Public Health* 97, no. 9 (2007): 1545–6.

82 Aysha Z. Akhtar et al., 'Health Professionals' Roles in Animal Agriculture, Climate Change, and Human Health', *American Journal of Preventive Medicine* 36, no. 2 (2009): 182–7.

83 Kate E. Jones et al., 'Global Trends in Emerging Infectious Diseases', *Nature* 451 (2008): 992.

84 Sonia Shah, *Pandemic: Tracking Contagions, from Cholera to Ebola and Beyond* (Sarah Crichton Books, 2016), 19.

85 'Paris's elite attended elaborate masquerade parties where, in denial and defiance of cholera's toll, they danced to "cholera waltzes", costumed as the ghoulish corpses many would soon become ... Every now and then, one of the revelers would rip off his mask, face purpled, and collapse. Cholera killed them so fast they went to their graves still clothed in their costumes.' Ibid., 43.

86 Although most biographical accounts agree that Hegel died of cholera, this interpretation is not unanimous. Terry Pinkard blames an 'upper gastrointestinal' disease for his death. Pinkard, *Hegel*, 616.

87 Hesiod, 'The Works and Days', in *Hesiod*, trans. Richmond Lattimore (University of Michigan Press, 1959), p. 29, lines 90–2.

88 See Marshall Sahlins 'The Original Affluent Society', in *Stone Age Economics* (Aldine-Atherton, 1972).

89 Hesiod, 'The Works and Days', p. 29, line 95.

90 Ibid., pp. 29–31, lines 101–3.

91 'The Bourgeois does not work for another. But he does not work for himself, taken as a biological entity, either. He works for himself taken as a "legal *person*", as a private *Property-owner*: he works for Property taken as such – i.e., Property that has now become *money*; he works for Capital.' Alexandre Kojève, *Introduction to the Reading of Hegel: Lectures on the Phenomenology of Spirit*, ed. Alan Bloom (Cornell University Press, 1980 [1947]), 65.

92 Marx, *Capital*, 1:255.

93 Karl Marx, 'Results of the Immediate Process of Production', in *Capital*, 1:1062.

94   Ibid., 1:988.

95   Quoted in Herzog, *Inventing the Market*, 55.

96   Marx, *Capital*, 1:254.

97   Max Horkheimer and Theodor W. Adorno, *Dialectic of Enlightenment: Philosophical Fragments* (Stanford University Press, 2002 [1944]), 26.

98   Ibid., 27.

99   Moishe Postone argued that capitalism could not stay still because 'increasing productivity increases the amount of use-values produced per unit of time, but results only in short-term increases in the magnitude of value created per unit of time. Once that productive increase becomes general, the magnitude of value falls to its base level. The result is a sort of treadmill dynamic.' See his 'Critique and Historical Transformation', *Historical Materialism* 12, no. 3 (2004): 59.

100 IPCC, *Global Warming of 1.5 °C: An IPCC Special Report* (2018), 148, 353, 354, 364.

101 John O'Neill, 'Who Won the Socialist Calculation Debate?', *History of Political Thought* 17, no. 3 (1996): 434.

102 Otto Neurath, 'Pseudorationalismus der Falsifikation', *Erkenntnis* 5 (1935): 353–65; Otto Neurath, 'The Problem of the Pleasure Maximum', in *Empiricism and Sociology*, ed. Marie Neurath and Robert S. Cohen (D. Reidel, 1973 [1912]), 113–22; Otto Neurath, 'Through War Economy to Economy in Kind', in *Empiricism and Sociology*, 146.

103 Neurath, 'Economy in Kind', 145.

104 Otto Neurath, 'Total Socialisation', in *Economic Writings, Selections 1904–1945*, ed. Thomas. E. Uebel and Robert S. Cohen (Springer, 2005 [1920]), 399.

105 Otto Neurath, 'What Is Meant by a Rational Economic Theory?', in *Unified Science*, ed. Brian McGuinness (D. Reidel, 1987 [1935]), 108.

106 Jordi Cat, 'Political Economy: Theory, Practice, and Philosophical Consequences', in *Stanford Encyclopedia of Philosophy* (Fall 2019 edition), ed. Edward N. Zalta, plato.stanford.edu.

107 Neurath, 'Economy in Kind', 131.

108 Ibid., 136, 141, 146–7.

109 Ibid., 142.

110 In the closely interconnected world of the Viennese intelligentsia, Mises knew Neurath personally as the two had attended the

economist Eugen Ritter von Böhm-Bawerk's seminar before the war. Even then, Mises despised Neurath for spouting 'nonsense' with 'fanatical fervor'. Bruce Caldwell, *Hayek's Challenge: An Intellectual Biography of F. A. Hayek* (University of Chicago Press, 2004), 114. For an overview of this debate, see Jessica Whyte, 'Calculation and Conflict', *South Atlantic Quarterly* 119, no. 1 (2020), 31–51.

111 Friedrich Hayek, 'The Nature and the History of the Problem', in *Collectivist Economic Planning*, ed. Friedrich Hayek (Routledge & Kegan Paul, 1935), 30.

112 Ludwig von Mises, *Economic Calculation in the Socialist Commonwealth* (Mises Institute, 1990 [1920]), 21n. In the original version of this text, Neurath was not even mentioned. Hayek later on explained that, indeed, it was Neurath's text that had 'provoked' Mises to write his 1920 essay. See Caldwell, *Hayek's Challenge*, 116.

113 Mises, *Economic Calculation*, 14.

114 'Both as an expression of recognition for the great service rendered by him and as a memento of the prime importance of sound economic accounting, a statue of Professor Mises ought to occupy an honourable place in the great hall of the Ministry of Socialisation or of the Central Planning Board of the socialist state.' Oskar Lange, 'On the Economic Theory of Socialism: Part One', *Review of Economic Studies* 4, no. 1 (1936): 53.

115 Thomas E. Uebel, 'Otto Neurath as an Austrian Economist: Behind the Scenes of the Early Socialist Calculation Debate', in *Otto Neurath's Economics in Context*, ed. Elisabeth Nemeth et al. (Springer, 2007), 41.

116 'A planning agency is likely to make widespread use of arithmetic, and indeed, if one wants to make localized decisions on the optimal use of resources by arithmetic means, then Mises' argument about the need to convert different products into some common denominator for purposes of calculation is quite correct. If, however, one wishes to perform global optimizations on the whole economy, other computational techniques, having much in common with the way nervous systems are thought to work, may be more appropriate, and these can in principle be performed without resort to arithmetic. Of course it would be anachronistic to fault Mises for failing to take into account developments in computer science which took place long after

he wrote.' Allin Cottrell and W. Paul Cockshott, 'Calculation, Complexity and Planning: The Socialist Calculation Debate Once Again', *Review of Political Economy* 5, no. 1 (1993): 79.

117 Friedrich Hayek, 'The Use of Knowledge in Society', *American Economic Review* 35, no. 4 (1945): 519.

118 Ibid., 526.

119 This term comes from Robert N. Proctor and Londa Schiebinger, eds, *Agnotology: The Making and Unmaking of Ignorance* (Stanford University Press, 2008).

120 Frank Knight, 'Some Fallacies in the Interpretation of Social Cost', *Quarterly Journal of Economics* 38, no. 4 (1924): 606.

121 Bruce Caldwell, 'F. A. Hayek and the Economic Calculus', *History of Political Economy* 48, no. 1 (2016): 161n.

122 Early on, Hayek complained: 'It seems that that skeleton in our cupboard, the "economic man", whom we have exorcised with prayer and fasting, has returned through the back door in the form of a quasi-omniscient individual.' See his 'Economics and Knowledge', *Economica* 4, no. 13 (1937): 45.

123 Philip Mirowski, *Never Let a Serious Crisis Go to Waste: How Neoliberalism Survived the Financial Meltdown* (Verso, 2014), 332.

124 Friedrich Hayek, 'The Trend of Economic Thinking', *Economica* 40 (1933): 123, 130. In later years, Hayek would add cybernetic and evolutionary insights to his economic analysis. Naomi Beck, *Hayek and the Evolution of Capitalism* (University of Chicago Press, 2018).

125 Hayek, 'The Trend of Economic Thinking', 123; Friedrich Hayek, 'The Pretence of Knowledge', lecture to the memory of Alfred Nobel, Stockholm, 11 December 1974, nobelprize.org.

126 H. Scott Gordon, 'The Economic Theory of a Common-Property Resource: The Fishery', *Journal of Political Economy* 62, no. 2 (1954): 124–42; Edward Nik-Khah, 'Neoliberal Pharmaceutical Science and the Chicago School of Economics', *Social Studies of Science* 44, no. 4 (2014): 489–517.

127 Mirowski, *Never Let a Serious Crisis Go to Waste*, 340.

128 John H. Dales, *Pollution, Property and Prices: An Essay in Policy-Making and Economics* (Edward Elgar, 2002 [1968]), 102–4. Similarly, Ronald Coase understands environmental problems as 'nuisances'. See Ronald Coase, 'The Problem of Social Cost', *Journal of Law & Economics* 3 (October 1960): 1–44.

129 Mises, *Economic Calculation*, 11.

130 Paul J. Crutzen, 'The Influence of Nitrogen Oxides on the Atmospheric Ozone Content', *Quarterly Journal of the Royal Meteorological Society* 96, no. 408 (1970): 320–5; Paul J. Crutzen, 'SST's: A Threat to the Earth's Ozone Shield', *Ambio* 1, no. 2 (1972): 41–51.

131 Mario J. Molina and F. Sherwood Rowland, 'Stratospheric Sink for Chlorofluoromethanes: Chlorine Atom-Catalysed Destruction of Ozone', *Nature* 249, no. 5460 (1974): 810–12.

132 This problem inspired Albert Einstein to join up with his friend Leo Szilard to make a much safer fridge – one that didn't use CFCs. Sam Kean, 'Einstein's Little-Known Passion Project? A Refrigerator', *Wired*, 23 July 2017, wired.com.

133 Susan Solomon, 'The Mystery of the Antarctic Ozone "Hole"', *Reviews of Geophysics* 26, no. 1 (1988): 131.

134 'An Undeniable Problem in Antarctica', *Understanding Science*, undsci.berkeley.edu.

135 Richard A. Kerr, 'Deep Chill Triggers Record Ozone Hole', *Science* 282, no. 5388 (1998): 391.

136 'An Undeniable Problem in Antarctica'.

137 Alan Robock, '20 Reasons Why Geoengineering May Be a Bad Idea', *Bulletin of the Atomic Scientists* 64, no. 2 (2008): 15–16.

138 David W. Keith et al., 'Solar Geoengineering Without Ozone Loss', *Proceedings of the National Academy of Sciences* 113, no. 52 (2016): 14910.

139 Daniel J. Cziczo et al., 'Unanticipated Side Effects of Stratospheric Albedo Modification Proposals Due to Aerosol Composition and Phase', *Scientific Reports* 9, no. 18825 (2019).

140 Kat Eschner, 'One Man Invented Two of the Deadliest Substances of the 20th Century', *Smithsonian Magazine*, 18 May 2017, smithsonianmag.com. Midgley's misadventures in invention eventually killed him: 'Later in life, he was struck by polio, writes *Encyclopedia Britannica*, and lost the use of his legs. Being of an inquiring mind, he invented a hoist mechanism to help him get in and out of bed. He died when he became tangled in the ropes and the device strangled him.'

141 Thomas J. Algeo and Stephen E. Scheckler, 'Terrestrial-Marine Teleconnections in the Devonian: Links Between the Evolution of Land Plants, Weathering Processes, and Marine Anoxic Events', *Philosophical Transactions of the Royal Society*

of London B: Biological Sciences 353, no. 1365 (1998): 113–30.

142 The first time Boulding used the idea was in Kenneth E. Boulding, 'The University, Society, and Arms Control', *Journal of Conflict Resolution* 7, no. 3 (1963): 458–63.

143 Kenneth E. Boulding, 'The Economics of the Coming Spaceship Earth', in *Environmental Quality Issues in a Growing Economy*, ed. Henry Jarrett (RFF Press, 1966), 9.

144 See for example Fred Scharmen, 'Jeff Bezos Dreams of a 1970s Future', *Bloomberg CityLab*, 13 May 2019, bloomberg.com.

145 Bruce Caldwell, 'Mont Pèlerin 1947', in *From the Past to the Future: Ideas and Actions for a Free Society*, 42, hoover.org.

146 David Keith, 'The Earth Is Not Yet an Artifact', *IEEE Technology and Society Magazine* 19, no. 4 (2000): 27.

147 Horkheimer and Adorno, *Dialectic of Enlightenment*, 1.

148 Neurath, 'Total Socialisation', 395.

149 'A huge whale hangs in the middle of the hall; but we do not learn how the "beard" is transformed into old-fashioned corsets, how the skin is transformed into shoes, or the fat into soap that finds its way to the dressing room of a beautiful woman. Nor do we learn how many whales are caught per annum, or how much whale bone, fat, and leather are produced by this means ... Human fortunes are connected with this exhibit – starving seamen, hungry families of fishermen in the north of Norway. And so, everything leads to men and society.' Otto Neurath, 'Museums of the Future', in *Empiricism and Sociology*, 219–20.

150 Otto Neurath, 'The Lost Wanderers of Descartes and the Auxiliary Motive (On the Psychology of Decision)', in *Philosophical Papers 1913–1946*, ed. Robert S. Cohen and Marie Neurath (D. Reidel, 1983 [1913]), 8.

151 Cameron Hu coined the term 'unbuilding' during our conversations about the book.

152 To finish the quotation: 'Let us recall that this Hegelian theme, among many others, was taken up by Marx. History properly so-called in which men ("classes") fight among themselves for recognition and fight against Nature by work, is called in Marx the "Realm of necessity" (*Reich der Notwendigkeit*); *beyond* (*jenseits*) is situated the "Realm of freedom" (*Reich der Freiheit*), in which men (usually recognizing one another without reservation) no longer fight, and work as little as possible

(Nature having been definitely mastered – that is, harmonized with Man).' Kojève, *Introduction to the Reading of Hegel*, 158–9n6.

153 Otto Neurath, 'A System of Socialisation', in *Economic Writings*, 345.

154 Otto Neurath, 'Anti-Spengler', in *Empiricism and Sociology*, 199.

2. A New Republic

1 Angela Davis, 'Social Justice in the Public University of California: Reflections and Strategies' (teach-in at the University of California Davis, 23 February 2012), video.ucdavis.edu. Lightly edited for clarity.

2 Plato, *The Republic*, trans. Paul Shorey, Loeb Classical Library (Harvard University Press, 1937 [1930]), 372b.

3 Ibid., 372c–d.

4 Ibid., 373b–d.

5 Ibid., 373e.

6 Plato, *Euthyphro. Apology. Crito. Phaedo. Phaedrus.*, trans. Harold N. Fowler, Loeb Classical Library 36 (Harvard University Press, 1914), 230d.

7 Thomas More, 'Utopia', in *Three Early Modern Utopias*, ed. Susan Bruce (Oxford University Press, 1999), 81.

8 Ibid., 22.

9 Ibid., 44.

10 Ibid., 23.

11 Ellen Meiksins Wood, *The Origin of Capitalism: A Longer View* (Verso, 2002), 109.

12 William Shakespeare, *The Life and Death of Richard II*, act 2, sc. 1.

13 Wood, *The Origin of Capitalism*, 152–6. The number of sheep jumped from 3 million to 4 million between 1500 and 1600. William Lazonick, 'Karl Marx and Enclosures in England', *Review of Radical Political Economics* 6, no. 2 (1974): 19. Over the longer arc of history, 1300 to 1850, livestock yields grew 400 per cent, a rate that surpassed increases in labour productivity (56 per cent) and grain harvests (120 per cent). Robert C. Allen, 'English and Welsh Agriculture, 1300–1850: Outputs,

Inputs, and Income' (working paper, Nuffield College, University of Oxford, 2006), 2, ora.ox.ac.uk.

14 Wood, *The Origin of Capitalism*, 153.

15 For reasons why the Tudors were wary of enclosures, see Lazonick, 'Karl Marx and Enclosures in England', 17.

16 The colonization of Ireland had already progressed by the Tudor period but deepened profoundly during Cromwell's reign. Donald Woodward, 'The Anglo-Irish Livestock Trade of the Seventeenth Century', *Irish Historical Studies* 18, no. 72 (1973): 489–523; Tiarnán Somhairle, 'Capital's First Colony? A Political Marxist Approach to Irish "Underdevelopment"', *Historical Materialism*, 5 February 2018, historicalmaterialism.org.

17 John Locke, *Second Treatise of Civil Government* (Prometheus Books, 1986 [1690]), 20.

18 Wood, *The Origin of Capitalism*, 162; More, 'Utopia', 63. More's Utopians 'count this the most just cause of war, when any people holdeth a piece of ground void and vacant to no good nor profitable use, keeping others from the use and possession of it.'

19 They would need $60–$150/tC. Sean Sweeney, 'Hard Facts about Coal: Why Trade Unions Should Rethink Their Support for Carbon Capture and Storage' (working paper, Trade Unions for Energy Democracy, New York, October 2015), 8, unionsforenergydemocracy.org. There is a list of all the cancelled CCS plants: MIT, 'Cancelled and Inactive Projects', Carbon & Sequestration Technologies, 30 September 2016, sequestration.mit.edu. The Institute for the Study of CCS at MIT, which devised the list, itself closed down in 2016.

20 See for example Carl-Friedrich Schleussner et al., 'Differential Climate Impacts for Policy-Relevant Limits to Global Warming: The Case of 1.5 C and 2 C', *Earth System Dynamics* 7, no. 2 (2016): 327–51; Richard J. Millar et al., 'Emission Budgets and Pathways Consistent with Limiting Warming to 1.5 C', *Nature Geoscience* 10, no. 10 (2017): 741–7.

21 For an overview of the concept, see Richard A. Houghton, 'Balancing the Global Carbon Budget', *Annual Review of Earth and Planetary Sciences* 35 (2007): 313–47.

22 UN Environment Programme, 'Cut Global Emissions by 7.6 Percent Every Year for Next Decade to Meet 1.5°C Paris Target – UN Report', press release, 26 November 2019, unenvironment.org; Simon Evans, 'Analysis: Coronavirus Set to Cause Largest

Ever Annual Fall in CO_2 Emissions', Carbon Brief, 9 April 2020, carbonbrief.org.

23 Zeke Hausfather, 'Analysis: Why the IPCC 1.5C Report Expanded the Carbon Budget', Carbon Brief, 8 October 2018, carbonbrief.org.

24 At the time, he called the technology 'biomass energy with CO_2 removal and permanent sequestration (BECS)'. Kenneth Möllersten, 'Opportunities for CO_2 Reductions and CO_2-Lean Energy Systems in Pulp and Paper Mills' (doctoral thesis, Royal Institute of Technology, Stockholm, 2002), kth.diva-portal.org.

25 Michael R. Obersteiner et al., *Managing Climate Risk* (IIASA interim report, 2001), 6, 16, pure.iiasa.ac.at. A much shorter version of the report was published in *Science* soon thereafter, where it has become a highly cited article: Michael R. Obersteiner et al., 'Managing Climate Risk', *Science* 294, no. 5543 (2001): 786–7.

26 David Keith, 'Sinks, Energy Crops and Land Use: Coherent Climate Policy Demands an Integrated Analysis of Biomass', *Climatic Change* 49 (2001): 7.

27 Ibid., 9.

28 One problem was the 'instability of carbon in biological reservoirs' compared with artificial sequestration like CCS. He also noted that 'the effectiveness of [natural] sinks is controversial, and depends critically on the timescale and management regime considered.' Ibid., 3.

29 Vera Heck et al., 'Biomass-Based Negative Emissions Difficult to Reconcile with Planetary Boundaries', *Nature Climate Change* 8, no. 2 (2018): 151–5. For the landmark planetary boundary study, see Will Steffen et al., 'Planetary Boundaries: Guiding Human Development on a Changing Planet', *Science* 347, no. 6623 (2015): 1259855.

30 Alister Doyle, 'Extracting Carbon from Nature Can Aid Climate but Will Be Costly: U.N.', Reuters, 26 March 2014, reuters.com.

31 Heck et al., 'Biomass-Based Negative Emissions', 153.

32 Anna Harper, 'Why BECCS Might Not Produce "Negative" Emissions After All', Carbon Brief, 14 August 2018, carbonbrief.org.

33 Keith, 'Sinks, Energy Crops and Land Use', 5.

34 Matteo Muratori et al., 'Global Economic Consequences of Deploying Bioenergy with Carbon Capture and Storage (BECCS)', *Environmental Research Letters* 11, no. 9 (2016): 4.

35 Philip Mirowski, *Never Let a Serious Crisis Go to Waste: How*

Neoliberalism Survived the Financial Meltdown (Verso, 2014), 332.

36 'Summary for Policymakers of IPCC Special Report on Global Warming of 1.5°C Approved by Governments', IPCC Newsroom, 8 October 2018, ipcc.ch.

37 Hansen has been arrested three times on the White House lawn (2010, 2011, 2013), and twice at protests elsewhere.

38 See James Hansen and Michael Shellenberger, 'The Climate Needs Nuclear Power', *Wall Street Journal*, 4 April 2019, wsj.com; James Hansen et al., 'Nuclear Power Paves the Only Viable Path Forward on Climate Change', *Guardian*, 3 December 2015, theguardian.com.

39 Hansen et al., 'Nuclear Power Paves the Only Viable Path'.

40 In the US in 2012, 15 per cent of fossil fuel use went to electricity, 52 per cent to liquid fuels, and 33 per cent to solid and gaseous fuels. Vaclav Smil, *Power Density: A Key to Understanding Energy Sources and Uses* (MIT Press, 2015), 246.

41 Spencer Wheatley, Benjamin Sovacool, and Didier Sornette, 'Of Disasters and Dragon Kings: A Statistical Analysis of Nuclear Power Incidents and Accidents', *Risk Analysis* 37, no. 1 (2017): 112.

42 Philip Ball, 'James Lovelock Reflects on Gaia's Legacy', *Nature*, 9 April 2014, nature.com; George Monbiot, 'Why Fukushima Made Me Stop Worrying and Love Nuclear Power', *Guardian*, 21 March 2011; George Monbiot, 'The Unpalatable Truth Is That the Anti-Nuclear Lobby Has Misled Us All', *Guardian*, 5 April 2011, theguardian.com. Hansen goes so far as to claim that nuclear power has in fact *saved* 1.8 million lives that would have been lost had that energy been provided by fossil fuels (a statistic soon cited by Shellenberger). Pushker A. Kharecha and James E. Hansen, 'Prevented Mortality and Greenhouse Gas Emissions from Historical and Projected Nuclear Power', *Environmental Science & Technology* 47, no. 9 (2013): 4889–95.

43 World Health Organization, 'World Health Organization Report Explains the Health Impacts of the World's Worst-Ever Civil Nuclear Accident', press release, 26 April 2006, who.int.

44 Ian Fairlie and David Sumner, *The Other Report on Chernobyl (TORCH)* (study for the European Parliament, Brussels, 2006); Kate Brown, *Manual for Survival: A Chernobyl Guide to the Future* (W. W. Norton, 2019), 311.

45 Debora Mackenzie, 'Caesium Fallout from Fukushima Rivals Chernobyl', *New Scientist*, 29 March 2011, newscientist.com; Tetsuji Imanaka, 'Comparison of Radioactivity Release and Contamination from the Fukushima and Chernobyl Nuclear Power Plant Accidents', in *Low-Dose Radiation Effects on Animals and Ecosystems: Long-Term Study on the Fukushima Nuclear Accident*, ed. Manabu Fukumoto (Springer, 2020), 257.

46 Jan Beyea et al., 'Accounting for Long-Term Doses in Worldwide Health Effects of the Fukushima Daiichi Nuclear Accident', *Energy & Environmental Science* 6, no. 3 (2013): 1042–5; Frank von Hippel, 'The Radiological and Psychological Consequences of the Fukushima Daiichi Accident', *Bulletin of the Atomic Scientists* 67, no. 5 (2011): 27–36.

47 'Accident Cleanup Costs Rising to 35–80 Trillion Yen in 40 Years', Japan Center for Economic Research, 3 July 2019, jcer.or.jp.

48 'Cleaning up Nuclear Waste Is an Obvious Task for Robots', *Economist*, 20 June 2019, economist.com.

49 Ben Dooley, Eimi Yamamitsu, and Makiko Inoue, 'Fukushima Nuclear Disaster Trial Ends with Acquittals of 3 Executives', *New York Times*, 19 September 2019, nytimes.com.

50 Associated Press, 'Japanese Power Company TEPCO Admits It Lied about Meltdown after Fukushima', CBC News, 21 June 2016, cbc.ca.

51 Indeed, assessments often omit these 'external' phases of the life cycle and focus only on generation. Keith Barnham, 'False Solution: Nuclear Power Is Not "Low Carbon"', *Ecologist*, 5 February 2015, theecologist.org.

52 Benjamin K. Sovacool, 'Valuing the Greenhouse Gas Emissions from Nuclear Power: A Critical Survey', *Energy Policy* 36, no. 8 (2008): 2941. Note that these measurements are CO_2 equivalents rather than CO_2 per se.

53 By contrast, the average coal-fired power plant emits around 1,050 gCO_2/kWh. Daniel Nugent and Benjamin K. Sovacool, 'Assessing the Lifecycle Greenhouse Gas Emissions from Solar PV and Wind Energy: A Critical Meta-Survey', *Energy Policy* 65 (2014): 229–44.

54 Simon Evans, 'Solar, Wind and Nuclear Have "Amazingly Low" Carbon Footprints, Study Finds', Carbon Brief, 8 December 2017, carbonbrief.org.

55 Terry Norgate, Nawshad Haque, and Paul Koltun, 'The Impact

of Uranium Ore Grade on the Greenhouse Gas Footprint of Nuclear Power', *Journal of Cleaner Production* 84 (2014): 365.

56 Ibid., Figs. 8–9. If one looks further ahead to only 0.001 per cent ore quality, then nuclear power's emissions would jump to 594 gCO_2/kWh – more than some fossil fuels. Ibid., 263.

57 See for example James Hansen, *Storms of My Grandchildren: The Truth about the Coming Climate Catastrophe and Our Last Chance to Save Humanity* (Bloomsbury, 2009), 201.

58 Frank von Hippel, 'Plutonium Programs in East Asia and Idaho Will Challenge the Biden Administration', *Bulletin of the Atomic Scientists*, 12 April 2021, thebulletin.org.

59 Frank von Hippel, 'Bill Gates' Bad Bet on Plutonium-Fueled Reactors', *Bulletin of the Atomic Scientists*, 22 March 2021.

60 Galina Raguzina, 'Holy Grail or Epic Fail? Russia Readies to Commission First Plutonium Breeder Against Uninspiring Global Track Record', Bellona, 4 August 2014, bellona.org; Thomas B. Cochran et al., 'It's Time to Give Up on Breeder Reactors', *Bulletin of the Atomic Scientists* 66, no. 3 (2010): 52.

61 Masa Takubo, 'Closing Japan's Monju Fast Breeder Reactor: The Possible Implications', *Bulletin of the Atomic Scientists* 73, no. 3 (2017): 182–7.

62 R. D. Kale, 'India's Fast Reactor Programme – A Review and Critical Assessment', *Progress in Nuclear Energy* 122 (2020): 103265; M. V. Ramana, 'A Fast Reactor at Any Cost: The Perverse Pursuit of Breeder Reactors in India', *Bulletin of the Atomic Scientists*, 3 November 2016.

63 S. Rajendran Pillai and M. V. Ramana, 'Breeder Reactors: A Possible Connection Between Metal Corrosion and Sodium Leaks', *Bulletin of the Atomic Scientists* 70, no. 3 (2014): 51–2.

64 The project's name referred not to the goal of transparency but to the idea that radiation was as omnipresent as sunshine. Here we also see the emergence of the modern notion of the 'environment'.

65 Joseph Masco, 'Bad Weather: On Planetary Crisis', *Social Studies of Science* 40, no. 1 (2010): 13–14. For the stolen baby bones, see Jacob Darwin Hamblin, *Arming Mother Nature: The Birth of Catastrophic Environmentalism* (Oxford University Press, 2013), 103.

66 Hamblin, *Arming Mother Nature*, 103; Sue Rabbitt Roff, 'Project Sunshine and the Slippery Slope: The Ethics of Tissue Sampling for Strontium-90', *Medicine, Conflict and Survival* 18, no. 3

(2002): 300, 304; Murray Campbell, 'Project Sunshine's Dark Secret', *Globe and Mail*, 6 June 2001, theglobeandmail.com.

67 Louise Zibold Reiss, 'Strontium-90 Absorption by Deciduous Teeth: Analysis of Teeth Provides a Practicable Method of Monitoring Strontium-90 Uptake by Human Populations', *Science* 134, no. 3491 (1961): 1669–73.

68 Rachel Carson, *Silent Spring* (Penguin, 2000 [1962]), 22.

69 The idea for such a study was sketched by Herman Kalckar, a biochemist at Johns Hopkins University. See his 'An International Milk Teeth Radiation Census', *Nature* 182, no. 4631 (1958): 283–4.

70 Reiss, 'Strontium-90 Absorption', 1669.

71 Harold L. Rosenthal, 'Accumulation of Environmental 90-Sr in Teeth of Children', in *Radiation Biology of the Fetal and Juvenile Mammal*, ed. Melvin R. Sikov and D. Dennis Mahlum (US Atomic Energy Commission, 1969), 163–71.

72 Barry Commoner and Ursula Franklin belonged to the Greater St. Louis Citizens' Committee on Nuclear Information.

73 Joseph J. Mangano et al., 'An Unexpected Rise in Strontium-90 in US Deciduous Teeth in the 1990s', *Science of the Total Environment* 317, nos. 1–3 (2003): 43. For a corroborative study see Jay M. Gould et al., 'Strontium-90 in Deciduous Teeth as a Factor in Early Childhood Cancer', *International Journal of Health Services* 30, no. 3 (2000): 515–39.

74 Karl Jacoby, *Crimes Against Nature: Squatters, Poachers, Thieves, and the Hidden History of American Conservation* (University of California Press, 2014); Jane Carruthers, *The Kruger National Park: A Social and Political History* (University of Natal Press, 1995).

75 For a history of the group, see Keith Makoto Woodhouse, *The Ecocentrists: A History of Radical Environmentalism* (Columbia University Press, 2018), 258–61.

76 Reed Noss, 'The Wildlands Project: Land Conservation Strategy', *Wild Earth*, special issue 'The Wildlands Project' (1992): 15. The difference between CPAWS' 50 per cent goal in 2005 and Noss' work in 1992 was that the Wildlands Network never adopted Noss' aim as official policy.

77 See Michael Soulé and Reed Noss, 'Rewilding and Biodiversity: Complementary Goals for Continental Conservation', *Wild Earth* 8 (1998): 18–28.

78 Edward O. Wilson, *The Diversity of Life* (Harvard University Press, 1992), 337. See also Harvey Locke, 'The International Movement to Protect Half the World: Origins, Scientific Foundations, and Policy Implications', in *Reference Module in Earth Systems and Environmental Sciences* (Elsevier, 2018).

79 See Edward O. Wilson, 'A Personal Brief for the Wildlands Project', *Wild Earth* 10, no. 1 (2000): 1–2. Wilson certainly knew his audience and wrote an alarmist essay on overpopulation. See also Harvey Locke, 'A Balanced Approach to Sharing North America', in the same issue.

80 Jenny Levison et al., *Apply the Brakes: Anti-immigrant Co-optation of the Environmental Movement* (Center for New Community, 2010), 3.

81 Ian Angus, 'Dave Foreman's Man Swarm: Defending Wildlife by Attacking Immigrants', review of *Man Swarm and the Killing of Wildlife*, by Dave Foreman, *Climate & Capitalism*, 25 April 2012, climateandcapitalism.com.

82 Indeed, Locke worked closely with Indigenous nations to expand nature preserves in Canada's Northwest Territories. Harvey Locke, 'Civil Society and Protected Areas: Lessons from Canada's Experience', *George Wright Forum* 26, no. 2 (2009): 101–28.

83 Dinitia Smith, 'Master Storyteller or Master Deceiver?', *New York Times*, 3 August 2002.

84 J. D. F. Jones, 'Van der Posture', *London Review of Books* 5, no. 2 (1983).

85 Harry Wels, *Securing Wilderness Landscapes in South Africa: Nick Steele, Private Wildlife Conservancies and Saving Rhinos* (Brill, 2015), 58.

86 For the WILD Foundation's conference programmes, see wild.org.

87 Malcolm Draper, 'Zen and the Art of Garden Province Maintenance: The Soft Intimacy of Hard Men in the Wilderness of KwaZulu-Natal, South Africa, 1952–1997', *Journal of Southern African Studies* 24, no. 4 (1998): 818.

88 Ibid., 817. Indeed, the new parks were under the jurisdiction of the IFP's conservation authority.

89 Adrian Guelke, 'Interpretations of Political Violence During South Africa's Transition', *Politikon* 27, no. 2 (2000): 241.

90 Zoo directors would visit Aspinall's estate in Kent to study his successful captive gorilla breeding programme. Aspinall impressed even zoologist and Nobel Prize winner Konrad Lorenz with his

'uncanny knack' with animals. Yet, he was a hardened Malthusian too. When told by Richard Nixon that a nuclear war might kill 200 million people, he thought that was too few. As a misogynist, he boasted of treating women 'with disdain' (especially if they were left-wing), and he likely abetted Lord Lucan's escape after the latter murdered his nanny. In 1997, Aspinall ran for Parliament under the banner of the Euro-sceptic Referendum Party. Malcolm Draper and Gerhard Maré, 'Going In: The Garden of England's Gaming Zookeeper and Zululand', *Journal of Southern African Studies* 29, no. 2 (2003): 551–69.

91   Stephen J. Gould, 'Cardboard Darwinism', *New York Review of Books*, 25 September 1986, 47–54.

92   Ed Douglas, 'Darwin's Natural Heir', *Guardian*, 16 February 2001; Edward O. Wilson, *Half-Earth: Our Planet's Fight for Life* (Liveright, 2016), 205.

93   Bram Büscher et al., 'Half-Earth or Whole Earth? Radical Ideas for Conservation, and Their Implications', *Oryx* 51, no. 3 (2017): 407–10.

94   Karl Marx, *Capital: A Critique of Political Economy*, vol. 1 (Penguin, 1976 [1867]), 494; Karl Marx, *Grundrisse: Foundations of the Critique of Political Economy* (Penguin, 1973), 612.

95   The ability to work, what Marx calls 'labour power', is different from other natural forces in that it is not merely exchanged as other commodities as equivalents (e.g., $100 of oil for $100 pork); instead, it provides the capitalist with the opportunity to capture surplus value. Marx, *Capital*, vol. 1, chapter 6.

96   Lazonick, 'Karl Marx and Enclosures in England', 16.

97   David Wykes, 'Robert Bakewell (1725–1795) of Dishley: Farmer and Livestock Improver', *Agricultural History Review* 52, no. 1 (2004): 39.

98   Karl Marx, *Capital: A Critique of Political Economy*, vol. 2 (Penguin, 1987 [1885]), 315.

99   Ibid.

100 Richard L. Hills, 'Sir Richard Arkwright and His Patent Granted in 1769', *Notes and Records of the Royal Society of London* 24, no. 2 (1970): 260.

101 Andreas Malm, 'The Origins of Fossil Capital: From Water to Steam in the British Cotton Industry', *Historical Materialism* 21, no. 1 (2013): 53.

102 Timothy Mitchell, *Carbon Democracy: Political Power in the Age of Oil* (Verso, 2011).

103 Kenneth Fish, *Living Factories: Biotechnology and the Unique Nature of Capitalism* (McGill-Queen's University Press, 2013), 141.

104 Ibid., 150.

105 Marx, *Capital*, 1:508 (translation amended).

106 Fish, *Living Factories*, 6.

107 For an excellent study of such tendencies in aquaculture, see Stefano B. Longo, Rebecca Clausen, and Brett Clark, *The Tragedy of the Commodity: Oceans, Fisheries, and Aquaculture* (Rutgers University Press, 2015).

108 M. J. Zuidhof et al., 'Growth, Efficiency, and Yield of Commercial Broilers from 1957, 1978, and 2005', *Poultry Science* 93, 12 (2014): 2980.

109 Don P. Blayney, *The Changing Landscape of U.S. Milk Production* (US Department of Agriculture Statistical Bulletin 978, 2002), appendix table 1, ers.usda.gov; 'Milk: Production per Cow by Year, US', United States Department of Agriculture, nass.usda.gov.

110 Hannah Ritchie and Max Roser, 'Land Use', *Our World in Data*, September 2019, ourworldindata.org.

111 Brian Machovina et al., 'Biodiversity Conservation: The Key Is Reducing Meat Consumption', *Science of the Total Environment* 536 (2015): 420.

112 Yinon M. Bar-On et al., 'The Biomass Distribution on Earth', *Proceedings of the National Academy of Sciences* 115, no. 25 (2018): 6508.

113 Robert Goodland and Jeff Anhang, 'Livestock and Climate Change', *World Watch* (November/December 2009): 11–12, awellfedworld.org.

114 Smil, *Power Density*.

115 Ibid., 113–14. A notable exception to this is Appalachian mountain-top removal, which has a power density 'well below 100 W/m²'.

116 Ibid., 146–7.

117 Chunhua Zhang et al., 'Disturbance-Induced Reduction of Biomass Carbon Sinks of China's Forests in Recent Years', *Environmental Research Letters* 10, no. 11 (2015): Table 1; Jingyun Fang et al., 'Changes in Forest Biomass Carbon Storage in

China Between 1949 and 1998', *Science* 292, no. 5525 (2001): 2320.

118 Joris P. G. M. Cromsigt et al., 'Trophic Rewilding as a Climate Change Mitigation Strategy?', *Philosophical Transactions of the Royal Society B: Biological Sciences* 373, no. 1761 (2018): 20170440.

119 Ibid., 3.

120 Ibid., 7.

121 Oswald J. Schmitz et al., 'Animating the Carbon Cycle', *Ecosystems* 17, no. 2 (2014): 348–9.

122 James W. Fourqurean et al., 'Seagrass Ecosystems as a Globally Significant Carbon Stock', *Nature Geoscience* 5, no. 7 (2012): 505.

123 Nicola Jones, 'How Growing Sea Plants Can Help Slow Ocean Acidification', *Yale Environment 360*, 12 July 2016, e360.yale. edu.

124 Joe Roman and James McCarthy, 'The Whale Pump: Marine Mammals Enhance Primary Productivity in a Coastal Basin', *PLoS ONE* 5, no. 10 (2010): e13255.

125 Andrew J. Pershing et al., 'The Impact of Whaling on the Ocean Carbon Cycle: Why Bigger Was Better', *PloS ONE* 5, no. 8 (2010): e12444.

126 World Wildlife Fund and Zoological Society of London, *Living Blue Planet Report: Species, Habitats and Human Well-Being* (WWF, 2015), 6; Rob Williams et al., 'Competing Conservation Objectives for Predators and Prey: Estimating Killer Whale Prey Requirements for Chinook Salmon', *PloS ONE* 6, no. 11 (2011): e26738; Elizabeth Pennisi, 'North Atlantic Right Whale Faces Extinction', *Science* 358, no. 6364 (2017): 703–4.

127 Wilson, *Half-Earth*, 136–51.

128 Vaclav Smil, *Harvesting the Biosphere: What We Have Taken from Nature* (MIT Press, 2013), 18–19.

129 Ulrich Kreidenweis et al., 'Afforestation to Mitigate Climate Change: Impacts on Food Prices under Consideration of Albedo Effects', *Environmental Research Letters* 11, no. 8 (2016): 085001.

130 Andy Skuce, '"We'd Have to Finish One New Facility Every Working Day for the Next 70 Years" – Why Carbon Capture Is No Panacea', *Bulletin of the Atomic Scientists*, 4 October 2016.

131 Karl-Heinz Erb et al., 'Exploring the Biophysical Option Space for Feeding the World Without Deforestation', *Nature Communications* 7 (2016): 11382.

132 Kreidenweis et al., 'Afforestation to Mitigate Climate Change', Table 1.

133 For a survey of the debate see Tomek de Ponti, Bert Rijk, and Martin K. van Ittersum, 'The Crop Yield Gap Between Organic and Conventional Agriculture', *Agricultural Systems* 108 (2012): 2.

134 David Pimentel et al., 'Environmental, Energetic, and Economic Comparisons of Organic and Conventional Farming Systems', *BioScience* 55, no. 7 (2005): 573–82.

135 Catherine Badgley and Ivette Perfecto, 'Can Organic Agriculture Feed the World?', *Renewable Agriculture and Food Systems* 22, no. 2 (2007): 81.

136 Janne Bengtsson, Johan Ahnström, and Ann-Christin Weibull, 'The Effects of Organic Agriculture on Biodiversity and Abundance: A Meta-Analysis', *Journal of Applied Ecology* 42, no. 2 (2005): 261–9.

137 Smil, *Power Density*, 247.

138 Ibid., 247–8.

139 Ibid., 246. His figures are 50 per cent displaced by PV (10 W/m^2) = 16,000 km², 25 per cent displaced by CSP (20 W/m^2) = 4,000 km², 25 per cent displaced by wind (50 W/m^2) = 1,600 km². One can quibble that Smil does not count the land between wind turbines because this can be cultivated, boosting their power density from around 10 W/m^2 to 50 W/m^2 in his calculations. However, even if one reduces wind power's contribution to a more likely 10 W/m^2, the system's land area would only increase to 28,000 km² (the system is a mix of wind and solar).

140 'After Many False Starts, Hydrogen Power Might Now Bear Fruit', *Economist*, 4 July 2020.

141 See Eberhard Jochem, ed., *Steps Toward a Sustainable Development: A White Book for R&D of Energy-Efficient Technologies* (CEPE/ETH Zurich, 2004). Two thousand watts as a rate of primary energy use per person equates to 17,500 kWh per year or 48 kWh per day.

142 Julia Wright, 'The Little-Studied Success Story of Post-crisis Food Security in Cuba', *International Journal of Cuban Studies* 4, no. 2 (2012): 131–2.

143 Ibid., 138; Sinan Koont, 'The Urban Agriculture of Havana', *Monthly Review* 60, no. 8 (2009): 50.

144 Sarah Boseley, 'Hard Times Behind Fall in Heart Disease and Diabetes in 1990s Cuba', *Guardian*, 9 April 2013.

145 This is the main argument of Emily Morris, 'Unexpected Cuba', *New Left Review* 1, no. 88 (2014): 5–45. She compares Cuba favourably with post-communist 'transitional' economies in Eastern Europe. The major crisis of the period, however, was Cuban epidemic neuropathy. In 1992, 30,000 people lost their eyesight due to nutritional deficiencies, but once the state had diagnosed the cause of the epidemic, it was able to respond quickly by distributing vitamin supplements via its robust primary care system. Rosaralis Santiesteban, 'In the Eye of the Cuban Epidemic Neuropathy Storm', interview by Christina Mills, *MEDICC Review* 13, no. 1 (2011): 10–15.

146 Elisa Botella-Rodriguez, 'Cuba's Inward-Looking Development Policies: Towards Sustainable Agriculture (1990–2008)', *Historia Agraria* 55 (2011): 160.

147 World Wide Fund for Nature, *Living Planet Report 2006* (World Wide Fund for Nature, 2006), 19.

148 Meghan E. Brown et al., 'Plant Pirates of the Caribbean: Is Cuba Sheltered by Its Revolutionary Economy?', *Frontiers in Ecology and the Environment* (2021); 'Cuba's Thriving Honey Business', *Economist*, 22 September 2018.

149 Stephen Milder, 'Between Grassroots Activism and Transnational Aspirations: Anti-Nuclear Protest from the Rhine Valley to the Bundestag, 1974–1983', *Historical Social Research/ Historische Sozialforschung* 39, no. 1 (147) (2014): 194.

150 Adrian Mehic, 'The Electoral Consequences of Nuclear Fallout: Evidence from Chernobyl' (working paper, Department of Economics, School of Economics and Management, Lund University, 2020), project.nek.lu.se.

151 Michael Shellenberger, 'The Real Reason They Hate Nuclear Is Because It Means We Don't Need Renewables', *Forbes*, 14 February 2019; Michael Shellenberger, 'Stop Letting Your Ridiculous Fears of Nuclear Waste Kill the Planet', *Forbes*, 19 June 2018, forbes.com.

152 James Hansen, 'Baby Lauren and the Kool-Aid', 29 July 2011, Columbia University, columbia.edu.

153 Michael Shellenberger, 'Democratic Presidential Candidates

Target Meat, Plastic Straws to Combat Climate Change, Reject Nuclear Power', interview by Dana Perino, *The Daily Briefing*, Fox News, 5 September 2019, video.foxnews.com.

3. Planning Half-Earth

1 Alan Bollard, *Economists at War: How a Handful of Economists Helped Win and Lose the World Wars* (Oxford University Press, 2020), 137–8.
2 Harrison E. Salisbury, *The 900 Days: The Siege of Leningrad* (Harper & Row, 1969), 412.
3 Bollard, *Economists at War*, 138.
4 Leonid V. Kantorovich, *The Best Use of Economic Resources* (Pergamon Press, 1965 [1959]), xviii.
5 Ibid., xxii.
6 Kantorovich's vision was popularized in the novel *Red Plenty*. See Francis Spufford, *Red Plenty* (Graywolf Press, 2012 [2010]).
7 Ibid., 254.
8 Leonid V. Kantorovich, 'Mathematical Methods of Organizing and Planning Production', *Management Science* 6, no. 4 (1960 [1939]).
9 Like calculus' twin births centuries earlier, linear programming was discovered a second time nearly a decade later in the United States. George Dantzig was a military planner who had been tasked with personnel and equipment procurement for the Air Force, which motivated his version of linear programming in 1947: the 'simplex' method. A year later, this tool rationalized efforts to relieve the Soviet siege of Berlin. Inspired by Dantzig's success, managers of private firms began to use linear programming to streamline their own operations. A notable difference between Dantzig and Kantorovich was that the former focused more on price, while the latter used in natura units. Robert Dorfman, 'The Discovery of Linear Programming', *Annals of the History of Computing* 6, no. 3 (1984): 283–95; Richard Cottle, Ellis Johnson, and Roger Wets, 'George B. Dantzig (1914–2005)', *Notices of the AMS* 54, no. 3 (2007): 344–62.
10 Leonid V. Kantorovich, 'My Journey in Science (Proposed Report to the Moscow Mathematical Society)', *Russian Math. Surveys* 42, no. 2 (1987): 233.

11 In the mid-century Soviet context, such a metric might be GDP or economic growth, though Neurath would surely disagree with optimizing quantities rooted in the 'pseudorational' world of money. In practice, this overall metric mattered little so long as constraints were well posed: if biodiversity and climate are respected while supplying everyone with enough energy and food, then whether we minimize energy use or land use matters only at the margins.

12 'At first talk was about countably-analytic calculators that had been acquired for the 1939 population census and after this were hardly used. Apparently these machines were first applied to numerical calculations by Professor Yanzhul from the Astronomical Institute in Leningrad. The possibilities for using these machines for other calculations were discussed at the seminar. They were very slow – the tabulator took half a second for an addition, and for multiplication between five and eight seconds. They were casually talking about initial developments in electronic computers and countably-analytic calculators constructed on the same principles (of type Mark I and Mark II in the USA)'. Kantorovich, 'My Journey in Science', 260.

13 Kantorovich, 'Mathematical Methods', 368.

14 Kantorovich, 'My Journey in Science', 259.

15 Bollard, *Economists at War*, 153.

16 See Eric Magnin and Nikolay Nenovsky, 'Calculating Without Money: Theories of In-Kind Accounting of Alexander Chayanov, Otto Neurath and the Early Soviet Experiences', *European Journal of the History of Economic Thought* 28, no. 3 (2020): 456–77.

17 Pareto condoned the new fascist regime but died a year after the March on Rome. Renato Cirillo, 'Was Vilfredo Pareto Really a "Precursor" of Fascism?', *American Journal of Economics and Sociology* 42, no. 2 (1983): 235–45. Neoclassical economics is a politically ambivalent tradition; despite its conservative reputation now, during the first half of the twentieth century many of its theorists were either liberal or socialist.

18 Kantorovich, 'My Journey in Science', 259.

19 Kantorovich contributed through his meticulous calculations of the atomic bomb's critical mass. Stanislav M. Menshikov, 'Topicality of Kantorovich's Economic Model', *Journal of Mathematical Sciences* 133 (2006): 1394.

20 Slava Gerovitch, *From Newspeak to Cyberspeak: A History of Soviet Cybernetics* (MIT Press, 2002), 269.

21 Adam E. Leeds, 'Dreams in Cybernetic Fugue: Cold War Techno-science, the Intelligentsia, and the Birth of Soviet Mathematical Economics', *Historical Studies in the Natural Sciences* 46, no. 5 (2016): 660. The vision of a half-human, half-machine building didn't work out as planned, and the building ended up being used more conventionally.

22 Cybernetics was a diffuse set of ideas about using computers to control complex systems, rather than to merely perform computations. 'Once cybernetics became en vogue, the general intellectual climate became yet more favourable for Kantorovich's ideas. Generally, the democratic tendencies of the Thaw and the widespread optimism about scientific progress, reinforced by the Sputnik euphoria, became two decisive factors in fostering the popularity of his notion of optimal planning.' Ivan Boldyrev and Till Düppe, 'Programming the USSR: Leonid V. Kantorovich in Context', *British Journal for the History of Science* 53, no. 2 (2020): 268.

23 Jenny Andersson and Eglė Rindzevičiūtė, 'The Political Life of Prediction: The Future as a Space of Scientific World Governance in the Cold War Era', *Les Cahiers européens de Sciences Po* 4 (2012): 13; Richard E. Ericson, 'The Growth and Marcescence of the "System for Optimal Functioning of the Economy" (SOFE)', *History of Political Economy* 51, no. S1 (2019): 165.

24 Edwin Bacon, 'Reconsidering Brezhnev', in *Brezhnev Reconsidered*, ed. Edwin Bacon and Mark Sandle (Palgrave Macmillan, 2002), 11–12; Abraham Katz, *The Politics of Economic Reform in the Soviet Union* (Praeger, 1972), 180–1.

25 János Kornai argues that the bureaucrats are bound together by (1) their values and shared belief in a noble purpose, (2) their 'resolve to retain power', (3) their prestige and relative privilege, and (4) self-discipline. János Kornai, *The Socialist System: The Political Economy of Communism* (Princeton University Press, 1992), 41–3. According to historians Ivan Boldyrev and Olessia Kirtchik, Soviet planning had changed little since its inception in the 1930s, remaining until the end a set of 'negotiations between different actors including Gosplan, ministries, and large industrial units competing for scarce resources'. Ivan Boldyrev and Olessia Kirtchik, 'The Cultures of Mathematical Economics in

the Postwar Soviet Union: More Than a Method, Less Than a Discipline', *Studies in History and Philosophy of Science* 63 (2017): 5.

26 Eglė Rindzevičiūtė, 'Toward a Joint Future Beyond the Iron Curtain: East–West Politics of Global Modelling', in *The Struggle for the Long-Term in Transnational Science and Politics: Forging the Future*, ed. Jenny Andersson and Eglė Rindzevičiūtė (Routledge, 2015), 130.

27 Boldyrev and Düppe, 'Programming the USSR'.

28 Leonid Kantorovich, 'Mathematics in Economics: Achievements, Difficulties, Perspectives', prize lecture given for the 1975 Sveriges Riksbank Prize in Economic Sciences in Memory of Alfred Nobel, nobelprize.org.

29 We are not the first to notice Kantorovich's and Neurath's frameworks as complementary. See Paul Cockshott, 'Calculation In-Natura, from Neurath to Kantorovich', University of Glasgow, 15 May 2008, dcs.gla.ac.uk.

30 This literature began to take off in 2009, when the planetary boundaries concept was coined in Johan Rockström et al., 'Planetary Boundaries: Exploring the Safe Operating Space for Humanity', *Ecology and Society* 14, no. 2 (2009), though the idea was around long before. Julia Steinberger's research group at the University of Leeds has done important work to further this research. See, for example, Daniel O'Neill et al., 'A Good Life for All Within Planetary Boundaries', *Nature Sustainability* 1, no. 2 (2018): 88–95. Another notable example is Kate Raworth, *Doughnut Economics: Seven Ways to Think Like a 21st-Century Economist* (Chelsea Green Publishing, 2017).

31 For an overview of the variety of IAMs used at present, see Alexandros Nikas, Haris Doukas, and Andreas Papandreou, 'A Detailed Overview and Consistent Classification of Climate-Economy Models', in *Understanding Risks and Uncertainties in Energy and Climate Policy: Multidisciplinary Methods and Tools for a Low Carbon Society*, ed. Haris Doukas, Alexandros Flamos, and Jenny Lieu (Springer, 2019), 1–54. Of particular interest to this book's readers will be computable general equilibrium models that undergird the scenarios used in IPCC climate reports.

32 For some discussions of IAMs, see for example Paul Parker et al., 'Progress in Integrated Assessment and Modelling',

Environmental Modelling & Software 17, no. 3 (2002): 209–17; Frank Ackerman et al., 'Limitations of Integrated Assessment Models of Climate Change', *Climatic Change* 95, nos. 3–4 (2009): 297–315; Lisette van Beek et al., 'Anticipating Futures Through Models: The Rise of Integrated Assessment Modelling in the Climate Science-Policy Interface Since 1970', *Global Environmental Change* 65 (2020): 102191.

33 One prominent *Nature* paper, which limits warming to 1.5°C without the need to use negative emissions technologies like BECCS, requires the dramatic curtailment of the meat industry, serious cuts to energy use, and rapid reforestation (policies parallel to what we advocate in this book). Detlef P. van Vuuren et al., 'Alternative Pathways to the 1.5°C Target Reduce the Need for Negative Emission Technologies', *Nature Climate Change* 8, no. 5 (2018): 391–7. See also Arnulf Grubler et al., 'A Low Energy Demand Scenario for Meeting the 1.5°C Target and Sustainable Development Goals Without Negative Emission Technologies', *Nature Energy* 3, no. 6 (2018): 515–27.

34 Otto Neurath, 'Through War Economy to Economy in Kind', in *Empiricism and Sociology*, ed. Marie Neurath and Robert S. Cohen (D. Reidel, 1973 [1919]), 151.

35 Ibid., 151, 154.

36 Ibid., 154.

37 See, for example, C. T. M. Clack et al., 'Linear Programming Techniques for Developing an Optimal Electrical System Including High-Voltage Direct-Current Transmission and Storage', *International Journal of Electrical Power & Energy Systems* 68 (2015): 103–14.

38 W. Paul Cockshott and Allin Cottrell, *Towards a New Socialism* (Spokesman, 1993). For a response to these 'labour money' schemes, see Jasper Bernes, 'The Test of Communism', March 2021, jasperbernesdotnet.files.wordpress.com.

39 Rockström et al., 'Planetary Boundaries'.

40 'Ambient (Outdoor) Air Pollution', World Health Organization, 2 May 2018, who.int.

41 The following paper estimates a global energy use of 149 EJ in 2050 (15.3 GJ/cap/yr), which is approximately 485 W. Joel Millward-Hopkins et al., 'Providing Decent Living with Minimum Energy: A Global Scenario', *Global Environmental Change* 65 (2020): 102168.

42 Emissions are from Peter Scarborough et al., 'Dietary Greenhouse Gas Emissions of Meat-Eaters, Fish-Eaters, Vegetarians and Vegans in the UK', *Climatic Change* 125, no. 2 (2014): 179–92. Land use is from Christian J. Peters et al., 'Carrying Capacity of U.S. Agricultural Land: Ten Diet Scenarios', *Elementa: Science of the Anthropocene* 4, no. 000116 (2016): Fig. 2. The 2.05 tonne figure is for 'medium meat-eaters'; high meat-eaters emit a whopping 2.62 tonnes.

43 For example, 'regenerative' agricultural practices – where farmers carefully manage their soil so that it absorbs more carbon from the atmosphere, using methods like no-till agriculture or the use of cover crops – could cut food emissions by as much as 70 per cent (though this field is notorious for its many implausible promises). Presumably other efficiency gains outside of food production could also help improve this figure, such as a transition to organic agriculture. 'Regenerative Agriculture: Good for Soil Health, but Limited Potential to Mitigate Climate Change', World Resources Institute, 12 May 2020, wri.org. When we run the model later in this chapter, we find that assuming these improvements on food emissions are necessary if we are to meet the objectives of our model, because if we assume food emissions for vegans are 1.05 tonnes per year, this alone exceeds the per capita emissions quota for the 1.5°C scenario for a 10-billion-person planet if we assume moderate climate sensitivity.

44 OECD/FAO, *OECD–FAO Agricultural Outlook 2018–2027* (OECD Publishing, 2018), Fig. 6.7, fao.org.

45 Summary report of the EAT-Lancet Commission, 'Food, Planet, Health: Healthy Diets from Sustainable Food Systems', 14.

46 See chapter two for our scepticism regarding nuclear. This power source could easily be added in a more sophisticated model.

47 Biofuel power densities are from Vaclav Smil, *Power Density: A Key to Understanding Energy Sources and Uses* (MIT Press, 2015), 226–9. Biofuels can have positive emissions due to the complexities of soil carbon. See Anna Harper, 'Why BECCS Might Not Produce "Negative" Emissions After All', Carbon Brief, 14 August 2018, carbonbrief.org.

48 Smil, *Power Density*, 227.

49 Emissions of natural gas are assumed to be 0.91 pounds of CO_2 per kWh, as reported at 'How Much Carbon Dioxide Is Produced Per Kilowatt-Hour of U.S. Electricity Generation?', US

Energy Information Administration, 15 December 2020, eia.gov. This converts to 3.61 kg CO_2/W, a figure which may be overly optimistic due to fugitive emissions and other upstream methane leaks. There is a wide range of power densities for natural gas electricity generation given in Smil's work, largely depending on the configuration of the plant. We chose 4,500 W/m², representing the upper range of power density for the Kawagoe plant in Japan's Mie prefecture, which was in the middle of estimates that ranged between 1,200 W/m² for some American plants to 20,000 W/m² for the Ravenwood plant in Queens, New York. Smil, *Power Density*, 142–3.

50 Smil, *Power Density*, 238–43.

51 David McDermott Hughes, 'To Save the Climate, Give Up the Demand for Constant Electricity', *Boston Review*, 1 October 2020, bostonreview.net.

52 See for example 'Energy Perspectives: Industrial and Transportation Sectors Lead Energy Use by Sector', US Energy Information Administration, 18 December 2012, eia.gov.

53 These energy-consuming sectors are included in every individual's 2,000-watt quota, in addition to the basic energy requirements of social goods, such as health care and education. In many statistical aggregations, agriculture is considered a part of industry.

54 'After Many False Starts, Hydrogen Power Might Now Bear Fruit', *Economist*, 4 July 2020, economist.com. We briefly discuss hydrogen in the second chapter.

55 Mark Z. Jacobson and Mark A. Delucchi, 'Providing All Global Energy with Wind, Water, and Solar Power, Part I: Technologies, Energy Resources, Quantities and Areas of Infrastructure, and Materials', *Energy Policy* 39, no. 3 (2011): 1154–69, Table 4.

56 Smil, *Power Density*, 244.

57 Ibid.

58 A 2016 study estimates that the amount of land used for biofuel globally was 413,000 km². Maria Cristina Rulli et al., 'The Water-Land-Food Nexus of First-Generation Biofuels', *Scientific Reports* 6, no. 22521 (2016): Table 1.

59 Otto Neurath, 'The Orchestration of the Sciences by the Encyclopedism of Logical Empiricism', *Philosophy and Phenomenological Research* 6, no. 4 (1946): 505.

60 Jordi Cat writes: '(1) Basic symbols must be self-evident, clear in themselves, representatives of a general concept or type; (2)

symbols must be independent of color; (3) the use of color is not regulated in general; (4) symbols must be drawn without perspective; (5) symbols must leave a vivid lasting impression on the mind; (6) symbols must be combinable; (7) a symbol can stand for a number of things, as a graphic unit, and a number of symbols then stand for a corresponding multiple number of things; (8) pictorial statistics are to be read from top left to bottom right like a book, except when comparing national statistics, on a geographical map; (9) combinations of symbols may form a unit of information like a story. Clearly, the pictorial language is semantically, syntactically and pragmatically limited and underdeveloped. In addition, its transmission relied on exemplars and training rather than an explicit theory.' Jordi Cat, 'Visual Education' (supplement to 'Otto Neurath'), in *Stanford Encyclopedia of Philosophy* (fall 2019 edition), ed. Edward N. Zalta (Stanford University, 1997), plato.stanford.edu.

61 Otto Neurath, *International Picture Language: The First Rules of ISOTYPE* (Kegan Paul, Trench, Trubner & Co., 1936), 22. See also Eve Blau, 'Isotype and Architecture in Red Vienna: The Modern Projects of Otto Neurath and Josef Frank', *Austrian Studies* 14 (2006): 227–59.

62 Wiener notes that 'we also wish to refer to the fact that the steering engines of a ship are indeed one of the earliest and best-developed forms of feedback mechanisms'. Norbert Wiener, *Cybernetics: Or Control and Communication in the Animal and the Machine*, 2nd ed. (MIT Press, 1961), 12.

63 Peter Galison, 'The Ontology of the Enemy: Norbert Wiener and the Cybernetic Vision', *Critical Inquiry* 21, no. 1 (1994): 228–66.

64 Ibid., 240.

65 Gerovitch, *From Newspeak to Cyberspeak*, 58–9.

66 Leeds, 'Dreams', 649.

67 Ibid., 650–2.

68 Benjamin Peters, 'Normalizing Soviet Cybernetics', *Information & Culture* 47, no. 2 (2012): 150, 160.

69 Leeds, 'Dreams', 665.

70 To simulate each variable in his model over time, say steel production, Kantorovich realized that he could treat it at every moment as a separate variable: 'Products of a particular type are differentiated with respect to the time period during which they are produced.' By applying a special set of rules to relate

these duplicated variables to one another, Kantorovich hacked the single-shot linear programming algorithm so that it could simulate change over time – it just had that many more variables to optimize. Leonid Kantorovich, 'A Dynamic Model of Optimum Planning', *Problems in Economics* 19, nos. 4–6 (1976 [1964]): 24–50.

71 Adam Leeds distinguishes between mathematical economists, who favoured optimization, and cyberneticians of various stripes, who favoured control. Leeds, 'Dreams', 664–5.

72 'The relationship between Kantorovich's institute and CEMI, however, has not been as close as one would expect. Kantorovich ... considered the expectations of CEMI to be exaggerated, specifically those relating to the so-called system of optimal functioning of the economy (SOFE) that had the ambition to indeed "programme the USSR".' Boldyrev and Düppe, 'Programming the USSR', 274.

73 Leonid V. Kantorovich et al., 'Toward the Wider Use of Optimizing Methods in the National Economy', *Problems in Economics* 29, no. 10 (1987 [1986]): 17.

74 Leeds, 'Dreams', 663.

75 Leeds summarizes this well, if perhaps making Kantorovich sound overly liberal: 'In sum, whereas the cyberneticians imagined a single well-*controlled* goal-oriented system, the economists imagined *calculating* parameters to steer another system, one external to the controllers: an evolving economy of independent agents.' Ibid., 664.

76 For the reader familiar with computer science: recent developments in linear programming have improved its computational complexity to current matrix multiplication time (see Michael B. Cohen, Yin Tat Lee, and Zhao Song, 'Solving Linear Programs in the Current Matrix Multiplication Time', arXiv.org, 18 October 2018). Furthermore, linear programming can make use of hardware acceleration made available by new architectures, such as Graphic Processing Units (GPUs), and in some cases can be sped up with sparse matrix calculations, or decomposed and parallelized. However, even this would not be enough to solve the problem. For a discussion on why planning the USSR in one giant linear programming model would be impossible, see Cosma Shalizi, 'In Soviet Union, Optimization Problem Solves *You*', *Crooked Timber*, 30 May 2012, crookedtimber.org.

77 Friedrich Hayek, 'The Use of Knowledge in Society', *American Economic Review* 35, no. 4 (1945): 526.

78 Eden Medina, *Cybernetic Revolutionaries: Technology and Politics in Allende's Chile* (MIT Press, 2014), 28.

79 Quoted in ibid., 26.

80 W. Ross Ashby, 'Requisite Variety and Its Implications for the Control of Complex Systems', in *Facets of Systems Science*, ed. George J. Klir (Springer, 1991), 405–17.

81 This exposition follows Medina's excellent explanation of Beer's viable systems model, which synthesizes several of Beer's writings. Medina, *Cybernetic Revolutionaries*, 34–9.

82 Ibid., 37.

83 Ibid.

84 Ibid., 38.

85 Ibid., 86.

86 Otto Neurath, 'A System of Socialisation', in *Economic Writings: Selections 1904–1945*, eds. Thomas E. Uebel and Robert S. Cohen (Springer, 2005), Fig. 5.

87 Henry Mauricio Ortiz Osorio and José Díaz Nafría, 'The Cybersyn Project as a Paradigm for Managing and Learning in Complexity', *Systema* 4, no. 2 (2016): 13.

88 Medina, *Cybernetic Revolutionaries*, 146–50.

89 Ibid., 81.

90 Ibid., 106.

91 Ibid. While CHECO was never fully operational, it did provide many simulations used by Chilean engineers.

92 Quoted in ibid., 107.

93 Ibid., 211.

95 Daniel Kuehn, '"We Can Get a Coup": Warren Nutter and the Overthrow of Salvador Allende', in *Research in the History of Economic Thought and Methodology: Including a Selection of Papers Presented at the 2019 ALAHPE Conference*, ed. Luca Fiorito, Scott Scheall, and Carlos Eduardo Suprinyak (Emerald Publishing, 2021), 151–86.

95 Soon after the 1980 constitution was implemented, Hayek told the Chilean newspaper *El Mercurio* that 'I prefer a liberal dictator to democratic government lacking liberalism'. For more on Hayek's relationship to the Pinochet regime, see Andrew Farrant and Edward McPhail, 'Can a Dictator Turn a Constitution into a Can-Opener? F. A. Hayek and the Alchemy of Transitional

Dictatorship in Chile', *Review of Political Economy* 26, no. 3 (2014): 331–48. On 26 October 2020, progressive forces won a referendum in Chile that abolished the 1980 constitution and set up a convention to draft a new one.

96 Diana Kurkovsky West, 'Cybernetics for the Command Economy: Foregrounding Entropy in Late Soviet Planning', *History of the Human Sciences* 33, no. 1 (2020): 44.

97 Olga Burmatova, *Optimizatsiia prostranstvennoi struktury TPK: ekologicheskii aspect* [Optimization of the spatial structure of TPC: The ecological aspect] (Nauka, 1983), 32, quoted in West, 'Command Economy', 47.

98 West, 'Command Economy', 48.

99 Burmatova, *Optimizatsiia*, 218, quoted in West, 'Command Economy', 48.

100 This enormous field is called 'data assimilation' and will be discussed briefly later in this section. The main algorithms used are Bayesian methods such as the Kalman filter (and its ensemble variant), 'variational' methods such as 4DVAR (ubiquitous in weather forecasting), and various other filters. For further reference, consult Sebastian Reich and Colin Cotter, *Probabilistic Forecasting and Bayesian Data Assimilation* (Cambridge University Press, 2015); Mark Asch, Marc Bocquet, and Maëlle Nodet, *Data Assimilation: Methods, Algorithms, and Applications* (SIAM, 2016); and Kody Law, Andrew Stuart, and Konstantinos Zygalakis, *Data Assimilation: A Mathematical Introduction* (Springer, 2015).

101 Rindzevičiūtė, 'Toward a Joint Future', 127–8.

102 Paul N. Edwards, *A Vast Machine: Computer Models, Climate Data, and the Politics of Global Warming* (MIT Press, 2010), 8.

103 Rasmus Benestad, 'Downscaling Climate Information', *Oxford Research Encyclopedia of Climate Science*, 7 July 2016, oxfordre.com.

104 See for example Reich and Cotter, *Probabilistic Forecasting*, 33–64.

105 Medina, *Cybernetic Revolutionaries*, 79.

106 Otto Neurath, 'Character and Course of Socialization', in *Empiricism and Sociology*, 140.

107 A highly active participant in debates over reform within Warsaw Pact nations, Kornai criticized both Kantorovich and Western neoclassical economists for their overly simplistic

understanding of real economies in his book *Anti-Equilibrium: On Economic Systems Theory and the Tasks of Research* (North-Holland Publishing Company, 1971).

108 Kornai, *The Socialist System*, 233–4 (emphasis in original).

109 So-called 'lemon socialism' can also occur in capitalist countries, where poorly run but vital or well-connected firms can always count on government bailouts in a crisis.

110 Robert Brenner, *The Economics of Global Turbulence: The Advanced Capitalist Economies from Long Boom to Long Downturn, 1945–2005* (Verso, 2006).

111 Also called 'Japanification' in the popular press. For an example of Summers' view, see Lawrence H. Summers, 'Accepting the Reality of Secular Stagnation', *Finance and Development* 57, no. 1 (2020): 17–19. For a Marxist perspective, see Aaron Benanav, *Automation and the Future of Work* (Verso, 2020).

112 Ibid., 569–70.

4. News from 2047

1 Karl Marx and Friedrich Engels, *The German Ideology* (Prometheus Books, 1998), 53.

2 Leonid Kantorovich, 'Mathematics in Economics: Achievements, Difficulties, Perspectives', lecture given for the 1975 Sveriges Riksbank (Bank of Sweden) Prize in Economic Sciences in Memory of Alfred Nobel, nobelprize.org.

3 Ron Scollon and Suzie Wong Scollon, 'The Axe Handle Academy: A Proposal for a Bioregional, Thematic Humanities Education', in *Lessons Taught, Lessons Learned: Teachers' Reflections on Schooling in Rural Alaska*, ed. Ray Barnhardt and J. Kelly Tonsmeire (Alaska State Department of Education, 1986), available at ankn.uaf.edu.

Epilogue

1 William Morris, *News from Nowhere* (Cambridge University Press, 1995 [1890]), 135.

2 Aaron Bastani, *Fully Automated Luxury Communism: A Manifesto* (Verso, 2019), 189. Bastani is an exception to the otherwise anti-environmentalist FALC collective in that he supports rewilding and abolishing the livestock industry.

3 For another melding of Hegel and Morris, see Rudolphus Teeuwen, 'An Epoch of Rest: Roland Barthes's "Neutral" and the Utopia of Weariness', *Cultural Critique* 80 (2012): 1–26.

4 J. Bruce Glasier, *William Morris and the Early Days of the Socialist Movement* (Longmans, Green, and Co., 1921), 150.

5 William Morris, review of *Looking Backward: 2000–1887*, by Edward Bellamy, *Commonweal* 5, no. 180 (1889): 194–5, marxists.org.

6 William Morris, 'The Aims of Art', in *Signs of Change: Seven Lectures Delivered on Various Occasions* (Reeves and Turner, 1888), 136.

7 Friedrich Hayek, 'The Trend in Economic Thinking', *Economica* 40 (1933): 123.

8 Friedrich Hayek, *The Counter-Revolution in Science: Studies on the Abuse of Reason* (Free Press, 1952), 88.

9 John Oswald, 'The Cry of Nature; Or, an Appeal to Mercy and to Justice, on Behalf of the Persecuted Animals', International Vegetarian Union, ivu.org.

10 Carol J. Adams, *The Sexual Politics of Meat: A Feminist-Vegetarian Critical Theory* (Continuum, 2010 [1990]), 268n6.

11 Percy Bysshe Shelley, *A Vindication of Natural Diet* (Offices of the Vegetarian Society, 1884 [1813]), 9, 20.

12 Edward Aveling and Eleanor Marx-Aveling, 'Shelley and Socialism', *To-Day*, April 1888, 103–16, marxists.org.

13 Adams, *The Sexual Politics of Meat*, 155.

14 Julia V. Douthwaite and Daniel Richter, 'The Frankenstein of the French Revolution: Nogaret's Automaton Tale of 1790', *European Romantic Review* 20, no. 3 (2009): 381–411.

15 Mary Shelley, *Frankenstein: Or the Modern Prometheus* (Routledge, 1888 [1818]), 77.

16 Fiona MacCarthy, review of *Edward Carpenter: A Life of Liberty and Love*, by Sheila Rowbotham, *Guardian*, 1 November 2008, theguardian.com.

17 Leah Leneman, 'The Awakened Instinct: Vegetarianism and the Women's Suffrage Movement in Britain', *Women's History Review* 6, no. 2 (1997): 271–87.

18 Karl Marx, *Capital: A Critique of Political Economy*, vol. 1 (Penguin, 1976 [1867]), 343.

19 George Orwell, *The Road to Wigan Pier* (Harcourt Brace, 1958 [1937]), 216, 222.

20 Ursula K. Le Guin, *A Wizard of Earthsea* (Houghton Mifflin, 2012 [1968]), 54.

21 Colin Burrow, 'It's Not Jung's, It's Mine', *London Review of Books* 43, no. 2 (2021): 12.

22 Ursula K. Le Guin, 'A Left-Handed Commencement Address', Mills College, 22 May 1983, americanrhetoric.com.

23 Ursula K. Le Guin, 'The Carrier Bag Theory of Fiction', in *The Ecocriticism Reader: Landmarks in Literary Ecology*, ed. Cheryll Glotfelty and Harold Fromm (University of Georgia Press, 1996), 154.